This text was created and copyrighted in 2020 by Luther M. Maddy, Ph.D.

This text has been reviewed and edited by Tammy Gammon, Ph.D.

All rights are retained. Reproduction of any part of this material is prohibited without the written permission of the author. The author makes no claims either expressed or implied as to the correctness or suitability of this material.

Microsoft® and Excel® are registered trademarks of Microsoft Corporation.

Published by:

Finish Line Press
Idaho, USA
www.FinishLinePress.com

For quantity discounts or to contact the author, visit: www.LutherMaddy.com.

Additional resources for this text are available on the author's website.

Table of Contents

Introduction: 1
Chapter 1: Statistics Defined 4
   Two Definitions of Statistics 4
   Two Categories of Statistics 5
      Descriptive Statistics 5
      Inferential Statistics 5
   Population vs Sample 5
      Why work with samples? 6
   Variables 7
      Qualitative Variables 7
      Quantitative Variables 7
   Four Quantitative Measurement Scales 9
      Nominal scale 9
      Ordinal scale 9
      Interval scale 9
      Ratio scale 10
   Chapter Highlights 11
Chapter 2: Descriptive Statistics – Frequencies 14
   Frequency Distributions of Categorical Variables 14
      Frequency table 14
      Bar or Column Chart 15
      Pie Chart 16
   Relative Frequencies 16
      Cumulative Relative Frequencies 17
   Frequency Distributions for Quantitative Data 17
      1.   Determine K, the number of classes (categories) 18
      2.   Determine the Interval or Width 19
   Chapter Highlights 22
Chapter 3: Descriptive statistics – Numerical Methods 24
   Measures of Central Tendency 24
      The Arithmetic Mean 24
      Parameters vs. Statistics 25
      Median – Locational Center Point 26
      Mode – The Most Popular Value 28
   Measures of Dispersion 28
      The range: 29
      Variance: 30
      Standard Deviation: 34
      The Empirical Rule 35
   Chapter Highlights 37
Chapter 4: Exploring Data 40
   Measures of Position 40
      Location vs. value 40
   Skewness 41

  Pearson's Coefficient of Skewness 42
  Software Skewness Computation 43
 Relationships between Two Variables 44
  A Scatter Plot 44
  Correlation Coefficient 45
 Chapter Highlights 47

Chapter 5: Exploring Probability Concepts 50
 Methods of Assigning Probability 50
  Classical Probability 51
  Relative Frequency or Empirical Probability 51
  Subjective Probability 51
 Computing Probabilities 52
  Important Terms 52
  Compliment Rule 52
  Rules of Addition 53
  Rules of Multiplication 54
 Counting Principles 56
  Multiplication Formula 56
  Combination Formula 56
  Permutation Formula 57
 Chapter Highlights 59

Chapter 6: Discrete Probability Distributions 62
 Probability Distributions 62
  Random Variables 62
  Discrete Probability Distributions 62
  Binomial Probability Distributions 65
  Poisson Probability Distributions 69
 Chapter Highlights 72

Chapter 7: Continuous Probability Distributions 76
 Uniform Distributions 76
 Normal Probability Distributions 78
  Standard Normal Distribution (Z scores) 79
 Chapter Highlights 86

Chapter 8: Sampling and Sampling Distributions 88
 Sampling methods 88
  Simple Random Sampling 88
  Systematic Random Sampling 88
 Sampling error 88
 Sampling Distribution of the Sample Mean 89
 The Central Limit Theorem 91
 Using a Sample Mean to Compute a Z score 92
 Standard Error of the Mean 94
 Chapter Highlights 95

Chapter 9: Confidence Intervals and Point Estimates 98
 Point Estimate 98
 Confidence Interval 98

- Margin of Error – Known Population Standard Deviation 98
- Margin of Error – Unknown Population Standard Deviation 100
- Z or T? 101
- Confidence Intervals for Proportions 102
- Choosing a sample size 103
- Chapter Highlights 105

Chapter 10: Hypothesis Testing: One Sample 108
- Five Step Hypothesis Testing Procedure 108
  1. State the null and alternate hypothesis 108
  2. Choose the level of significance 110
  3. Determine the test statistic 110
  4. Locate the critical value (decision rule) 110
  5. Take a sample, compute the test statistic and make a decision 111
- One Sample Hypothesis Test – Known Population Standard Deviation 112
- One Sample Hypothesis Test – Unknown Population Standard Deviation 115
- Hypothesis Test – One Proportion 117
- Chapter Highlights 120

Chapter 11: Two sample hypothesis testing 122
- Two Sample Hypothesis Testing – Known Population Standard Deviation 122
- Two Sample– Unknown Standard Deviation 124
- Two Sample Hypothesis Test – Proportions 127
- Two Sample Hypothesis Test – Dependent Samples 128
- Chapter Highlights 133

Chapter 12: ANOVA (Analysis of Variance) 136
- Comparing two population variances 136
- Completing an ANOVA table – three or more means 138
- Chapter Highlights 145

Chapter 13: Correlation and Linear Regression 148
- Dependent and Independent Variables 148
- Correlation Coefficient 148
- Regression Analysis 150
  - Slope 150
  - Intercept 150
  - Regression formula 151
  - Standard Error of the estimate 153
  - Coefficient of Determination 153
- Multiple Regression 153
  - Dummy Variables 154
  - The multiple regression equation 156
  - Evaluating a Multiple Regression Output 156
- Chapter Highlights 160

Chapter 14: Goodness of Fit Tests 164
- Goodness of Fit – Equal Expected Frequencies 164
- Goodness of Fit – Unequal Expected Frequencies 166
- Chapter Highlights 168

Index 169

Introduction:

Simply mentioning the word "mathematics" strikes fear into the hearts of many people. Throw the word "statistics" into the sentence and the reaction is even worse. If you have picked up this book, you may be struggling with a statistics course and perhaps even fear you will not be successful learning the topic. You may not even care about math or statistics, but just need to pass a college course, earn a degree, and get on with your life, hopefully putting statistics and math, behind you forever.

As a college business professor, when someone asks me what course I most enjoy teaching, I respond "statistics" with genuine enthusiasm. I truly enjoy teaching introductory statistics to people who hate math!

While I like numbers, I am not a mathematician. I am certainly not a statistician either. I struggled learning statistics in college, just like many of my students. The lowest grade I ever received in my undergraduate coursework was in statistics. The last day of my statistics course was one of the happiest days of my life.

I took additional statistical analysis courses in my graduate programs, many years after completing my undergraduate degree. As before, I struggled with the topic, but I made it through.

As a new professor I was assigned to teach introductory statistics courses. My first few times teaching statistics were also a struggle.

With a few semesters under my belt, I soon realized I could explain statistical concepts in a way many "math fearers" could understand. My classes were nearly always filled. I uploaded some videos on YouTube and viewers commented that they greatly appreciated my instruction. Some commented that my videos kept them from failing. I have compiled this guide to help students survive and even thrive in their statistics course, which can sometimes be a very difficult course to get through.

I designed this book to help those who may be struggling, as I did, to understand statistical concepts. This is not a text preparation book, but it may help raise your scores. This book is not designed to be comprehensive and it may skip some very important concepts. Consider this book a visit to a tutor who can explain what you may have already heard in a manner that makes more sense, or at least somewhat differently.

I wish you the best as you work your way through introductory statistics. This text in conjunction with the other resources you have for your class will, hopefully help you succeed. Who knows, you may even find yourself teaching statistics to others someday.

Luther M. Maddy III, Ph.D.

Chapter 1: Statistics Defined

This chapter will cover:

Defining Statistics

Understanding Populations and Samples

Descriptive and Inferential Statistics

Nominal, Ordinal, Interval, and Ratio Measurement Scales

Chapter 1: Statistics Defined

A statistics guide or textbook should start by defining "statistics".
So, what is statistics? Or, should the question be what are statistics?

Both questions are correct. And so, the confusion begins right on the first page.

Here is an attempt to clear up the definition.

Two Definitions of Statistics

1. Statistics **IS** the science (branch of mathematics) that deals with collecting, organizing, analyzing, presenting, and interpreting data (mass quantities of numbers), most often to assist decision making processes.

2. Statistics **ARE** organized, analyzed, presented, or interpreted quantitative data (numbers).

*So, you use **statistics** to produce **statistics**.*

As you move through this book, this will make more sense. Trust me for now.

Another definition I like: *Statistics is the process of producing useful information from raw data.*

58	26.23	13	38	53
17	3.02	15	47	22
18	3.25	8	40	44
19	2.36	12	21	37
20	2.33	9	50	21
20	0.71	11	41	57
20	2.92	16	50	46
23	0.71	9	50	50
24	0.62	22	50	40

Figure 1: Raw data

Number of patients surveyed	9
Average age of respondent	24.333
Average days sick per respondent	12.778
Average days in the hospital per respondent	4.683

Figure 2: Summarized data (information)

Which figure above makes more sense? Of course, it is Figure 2 with the summarized information. Raw data is important, but it becomes more useful after we use statistics, *(the process of summarizing, and presenting data)* to display some statistics *(organized and summarized information)* based on that raw data.

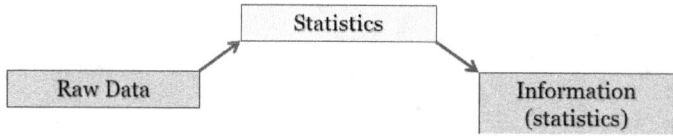

Figure 3: The Statistical Process

Two Categories of Statistics

While they are both considered statistics, there are two types of statistical information you will compute: descriptive and inferential.

DESCRIPTIVE STATISTICS

As the term implies, descriptive statistics describe the data. A descriptive statistic may be a simple as counting the number of people in a classroom. Descriptive statistics can move up the scale of difficulty and include computations like average, variance, standard deviation, skewness, and beyond. You will become familiar with these terms if they are new to you now. The important point is to understand that <u>descriptive statistics describes the data</u>.

INFERENTIAL STATISTICS

As this term implies, inferential statistics result in inferences, decisions, predictions or estimates. Think about an election. Before the election, if you watch or listen to the news, you would probably hear pundits predicting the likely winner. These pundits are making their predictions based on statistics. You may even hear them throw out percentages such as, "there is a 95% chance that Candidate A is going to win."

Inferential statistics are an important part of introductory statistics. Descriptive statistics are useful, but they simply describe what is. Inferential statistics try to predict, or infer, what might be, based on statistical information.

Researchers (and students in their statistics courses) use inferential statistics to make inferences, predictions, decisions, or estimates for *populations* based on data we or others have gathered from *samples*.

It may be time to define two of the terms I just used.

Population vs Sample

A **population** includes every person, or object in whatever group is being studied. For example, a population might be every car on a particular car lot or every student attending a college. Populations can be quite large, such as every person residing in the United States, but they do not have to be large. For instance, we might consider every student attending specific statistics course a population.

Figure 4: A population

Now, back to inferential statistics. Let's assume you want to know how many "happy faces" there are in the example population in the illustration. Answering that question is easy. Just count them. But this gets trickier.

Assume you want to know the makeup of this population. More specifically, you want to know the ratio of happy faces to the entire population. Or, you need to find the percentage of the entire population that are happy faces.

Answering these two questions requires you to look at (*survey*) every shape in the population in Figure 4. You will need to check every shape in the population to see if it is a happy face. You also need to examine and count every shape so you can find the total number of shapes in the population as well as the number of happy faces. The answer that you should have computed by now is 6/15 or 40%.

A **sample** is a representative portion, a subset of a population. Figure 5 illustrates a sample of three symbols from the population of 15 symbols. Inferential statistics, as we stated earlier, makes inferences, predictions, decisions, or estimates for populations based on data gathered from samples.

Using the sample from the illustration, the percentage of happy faces is 1/3, or 33.3%. Based on this sample we might infer that the ratio in the population is also 1/3. In this case, the percentage of happy faces in the sample is not quite the same as the percentage in the population. We will deal with that issue later.

Figure 5: A sample of a population

WHY WORK WITH SAMPLES?

As we just saw in the previous paragraph, a sample may lead to an incorrect assumption (inference) about the population. If a sample may not be an accurate representation, why take a sample instead of surveying the population?

First, let's briefly address the accuracy issue. You will learn in later chapters that samples can and do provide an accurate way to make inferences about populations. You will also learn how to deal with potential inaccuracies. For now, we will assume that samples can be accurate representations of a population.

Reasons to use samples instead of the entire population vary but include:

- Surveying the entire population would be expensive and time consuming.
    Consider a population as large as every resident in the United States. Surveying every individual in the country would be expensive and time consuming. The United States only attempts to do this once every ten years with the US Census.

- Sometimes surveying or testing is destructive.
    Suppose you oversee quality for a candy bar manufacturer. To test the quality, you have to unwrap the bar and taste it. Selling candy bars would be very difficult to sell your candy if every candy bar is unwrapped and has your teeth marks in it. Sampling is a better choice.

Variables

Surveying or testing a sample or a population involves asking questions, testing, or measuring. In your survey or test, you are interested in certain values or characteristics. Those values or characteristics are **variables**. For example, if you are studying basketball teams, you may decide to measure the height of each player. In this case, *height* would be considered a variable. There are two broad categories of variables: *quantitative* and *qualitative*.

QUALITATIVE VARIABLES

Qualitative variables are nonnumeric

Assume you still work at the candy company received a promotion. Now you are doing market research for your employer. You are a basketball fan, so you decide to survey the players on your college's team about their favorite candy bar. Their favorite candy bar is a **qualitative** variable because it is not a number.

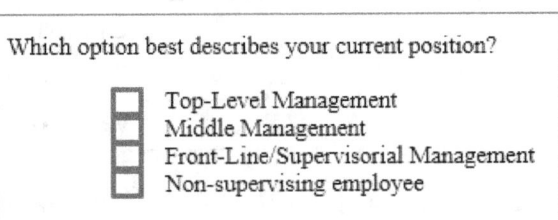

Figure 6: A survey question - qualitative

Continuing with your survey, you ask each player to explain why a particular candy bar is his or her favorite. The answer to your "why" question is also a variable. If the player was particularly passionate about their favorite candy bar, this variable could be several paragraphs of text. Not unimportant, just lengthy.

QUANTITATIVE VARIABLES

Quantitative variables are numeric. These might be measurements of height, weight, test scores, or income. Quantitative variables can also be counts such as how many children are in a family, how many cars are in a parking lot, or how many students are absent from a class.

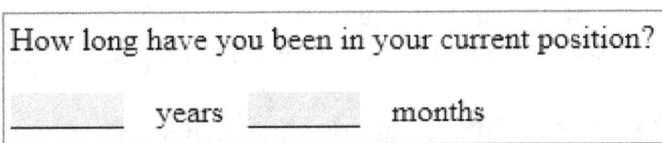

Figure 7: A survey question - quantitative

Since *statistics*, (i.e. the process) deals with numbers, (i.e. *statistics*) it may be obvious that we will be dealing with quantitative variables more often than qualitative variables.

For instance, if a survey variable was "favorite color," how would you compute an average of the sample's favorite color?

Using Qualitative Variables in Statistics

Choose your favorite color from the list below:		
Your Choice	Color	Asssigned value
	Red	1
	Blue	2
	Green	3
	Purple	4
	Yellow	5

Figure 8: Categorical data as quantitative

Do not give up on qualitative variables entirely because we can use them in statistics. However, we must first convert them into quantitative variables. For example, we could convert the choice of favorite color into a number. You could assign a value of 1 to red, 2 to blue and so on. You can also do something similar with the variable "gender" or other qualitative variables. Qualitative variables that can be converted into numerical data are called *categorical variables*.

While some qualitative variables can be converted to numbers, others are more difficult. Think about the variable, "Why do you like your favorite candy bar?" Since the answers can vary greatly, it would be very difficult to convert it into a numeric value. Qualitative variables like the "why" question are useful, just not in statistical analysis. (There are tools and techniques to help make sense of and analyze qualitative variables in research, but that is not something for this course. You will learn about these tools in a qualitative research methods course.)

Types of Quantitative Variables

As you have probably guessed, the statistical process is concerned with quantitative variables. So, we now need to discuss two main types of quantitative variables and four measurement scales we use to classify them.

Discrete variables result from counting. The number of students enrolled in a class, the number of devices connected to the internet in a home, and the number of patients in a hospital are examples of discrete variables.

Continuous variables are measured rather than counted and can assume any value for the item being measured. The weight of patients entering a hospital, the amount of time it takes students to complete an exam, and the amount of gasoline in your tank are all examples of continuous variables.

Four Quantitative Measurement Scales

Quantitative variables are classified into one of four measurement scales. The scale we use controls what types of statistical analysis we can perform with that variable.

To introduce this concept, consider two variables, the heights of basketball players on a team and numeric values assigned to a favorite color choice. Counting how many people like red and how many like blue would be useful. However, computing an average of the sample's favorite color would be worthless. Blue and a half, makes no sense. On the other hand, the average height of a basketball player is very useful. The scale a variable uses limits the statistical computations that are useful. The four scales used for quantitative variable are **nominal**, **ordinal**, **interval**, and **ratio**.

NOMINAL SCALE

A nominal scale is used for data than can be categorized and assigned values, but those values do not equate to relative worth or order. Again, think about the variable for favorite color assigned values. While you may prefer blue, that color is not more valuable or better than any other color. So, assigning red a value of 1 and blue a value of 2 does not mean that either has more relative value than the other. Beyond counting and computing percentages, statistical analysis of data that uses a nominal scale is very limited.

ORDINAL SCALE

Data that uses an ordinal scale can be ranked or ordered in a meaningful way. However, the differences between items in the ranked list either cannot be determined or are not meaningful. For example, you can list your favorite, next favorite and third favorite ice cream flavors as 1, 2, and 3. The order is important, you cannot numerically determine how much you like flavor #1 more than flavor #3. As with nominal scale data, analysis is limited to counting and percentages with variables using the ordinal scale.

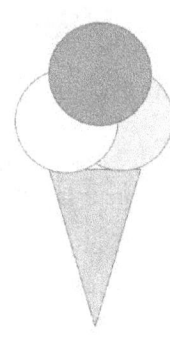

INTERVAL SCALE

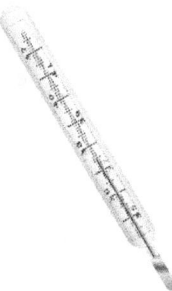

The interval scale is used for data with meaningful and measurable differences between values. Many additional statistical computations can be performed with data using the interval scale than can be performed with data using the nominal or interval scales. One of the identifying attributes of the interval scale is that zero does not really indicate a null value. The classic example of the interval scale is temperature. Neither 0°C or 0°F equates to the absence of temperature. Some clothing items, such as ladies' dresses and infant shoes also use the interval scale. A shoe size of zero does not mean the absence of any shoes.

RATIO SCALE

The major attribute that sets the ratio scale apart from the interval scale is that zero does mean zero. Both discrete and continuous variables can be measured using the ratio scale. One example of a discrete variable measured with the ratio scale might be the number of students in class on a particular day. In this case, zero means no students attended. A continuous variable using the ratio scale might be the amount of income taxes paid or the amount of gas in your tank.

Chapter Highlights

- Statistics is the process of collecting, organizing, analyzing, presenting, and interpreting data.

- Descriptive statistics summarizes and describes data so that it becomes information.

- Inferential statistics attempts to estimate, generalize, or predict something about a population.

- A population includes all the elements or individuals in the study group

- A sample is a subset of the population

- Why use samples?
    - Less costly and more practical than surveying every member of the population.
    - Some tests damage or destroy the item tested

- Variable: characteristic of interest that is tested, measured, or collected

- A qualitative variable is nonnumeric variable such as, favorite color or gender.
    - A categorical variable is a qualitative variable that can be assigned a value,

- A quantitative variable is a numeric variable.

- Quantitative variable measurement scales
    - Nominal: values do not equate to relative worth or order.

    - Ordinal: values ranked in a meaningful way but the differences between values cannot be determined with a numeric value.

    - Interval: meaningful and measurable differences between values but zero is not a null value.

    - Ratio: meaningful and measurable differences between values and zero is a null value.

Surviving Statistics

Chapter 2: Descriptive Statistics – Frequencies

This chapter will cover:
- **Frequency Distributions and Tables**
- **Bar, Column, and Pie Charts**
- **Relative and Cumulative Frequencies**
- **Frequency Distributions for Quantitative Data**

Chapter 2: Descriptive Statistics – Frequencies

One of the first questions to ask about collected data, and one of the easiest to answer, is "How many?" For example, "How many of the people we surveyed like our company's candy bar better than our competitors when we allowed them to choose?" We can answer these questions by counting. The result of our counts become the frequency of occurrence. We can display the number that chose each candy bar in frequency distributions as tables, charts, or graphs.

Frequency Distributions of Categorical Variables

If the data you are using has categorical variables, frequency distributions summarize the counts of each category. For example, with the favorite color variable, a frequency distribution would show the number of each person surveyed (observations) who selected, red, blue, green, or other color.

FREQUENCY TABLE

The frequency table in Figure 9 displays every possible choice for the variable and then the count for each. In all, 157 individuals were asked to select their favorite color from a set list of choices. The frequency table reports or describes, how many selected each color. When you create a frequency table, you should sort the category choices to make it easier to read.

Frequency Table	
Color choice	**# choosing**
Blue	31
Green	29
Orange	30
Red	42
Yellow	25
Total	157

Figure 9: Frequency Table

Excel as a Calculator

As we progress through this text, we will eventually be doing mathematical computations. Using a calculator to perform those computations can be helpful. My preferred calculator is Excel. As we get into those computations, I will discuss the math behind the formula so you can compute it manually or with a calculator. I will also show you how to do the computation in Excel longhand, and when one is available, with an Excel shortcut. I will also assume you have some basic background in Excel. If you have not worked with Excel before, there are many tutorials and resources available on the web, including those I have created which are on my website and for sale at online booksellers.

You can use Excel's CountIf() to count the frequency of occurrence of one categorical value in a large list. You can see this illustrated in Figure 10.

	A	B	C	D	E	F	G
1	ID	First	Last	Gender	Pay	Residence	Favorite color
2	5344	C	Alcantar	Female	$ 11.90	Nampa	Blue
3	5927	T	Allen	Female	$ 10.65	Boise	Red
4	5447	S	Allison	Female	$ 13.65	Caldwell	Green
5	5356	K	Alverson	Female	$ 12.20	Nampa	Yellow
6	4848	B	Amos	Female	$ 13.22	Meridian	Orange
7	5854	N	Anderson	Female	$ 16.15	Meridian	Green
8	5393	J	Anderson	Male	$ 12.25	Nampa	Yellow
9	5312	E	Anderson	Female	$ 11.30	Caldwell	Red
10	5239	N	Angel	Female	$ 15.36	Boise	Green
11	5885	M	Araiza	Male	$ 26.22	Caldwell	Blue
12	5886	M	Araiza	Female	$ 17.15	Caldwell	Red
13	5728	C	Arens	Female	$ 17.23	Boise	Green
14	5144	P	Armiger	Female	$ 16.48	Boise	Yellow
15	5733	S	Arnold	Female	$ 17.23	Meridian	Orange
16	5874	G	Au	Male	$ 25.78	Boise	Red
17	5558	D	Austin	Male	$ 15.05	Boise	Red
18	5549	J	Bair	Female	$ 15.30	Nampa	Blue
19	5986	D	Baker	Male	$ 27.10	Meridian	Blue
20	5394	K	Barnes	Female	$ 15.00	Nampa	Blue
21	5339	S	Barton	Female	$ 11.80	Caldwell	Orange

Frequency Table	
Color choice	**# choosing**
Blue	=COUNTIF(G2:G159,I8)
Green	COUNTIF(range, criteria)
Orange	30
Red	42
Yellow	25
Total	132

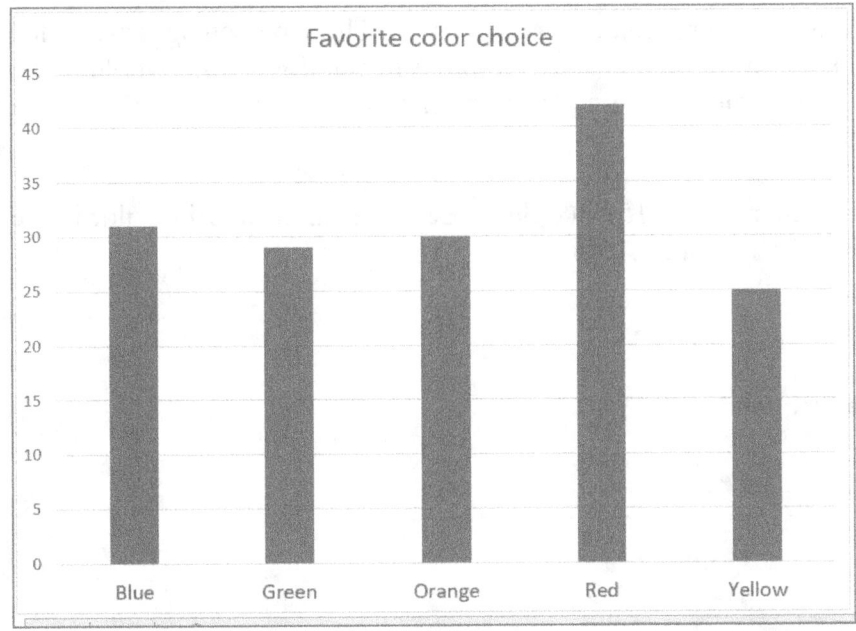

Figure 10: Using the CountIf() function to create a frequency table

BAR OR COLUMN CHART

In addition to a table, frequency distributions can be displayed as a bar or column chart which provide a graphical representation of the frequencies.

Figure 11: A column chart that displays frequencies

You can easily create bar or column charts in Excel or other software.

PIE CHART

A pie chart is also an effective way to represent frequencies graphically. A pie chart has the advantage of showing how each color choice relates to the choices of all the colors. In other words, a pie chart shows the relative frequency of each choice (color).

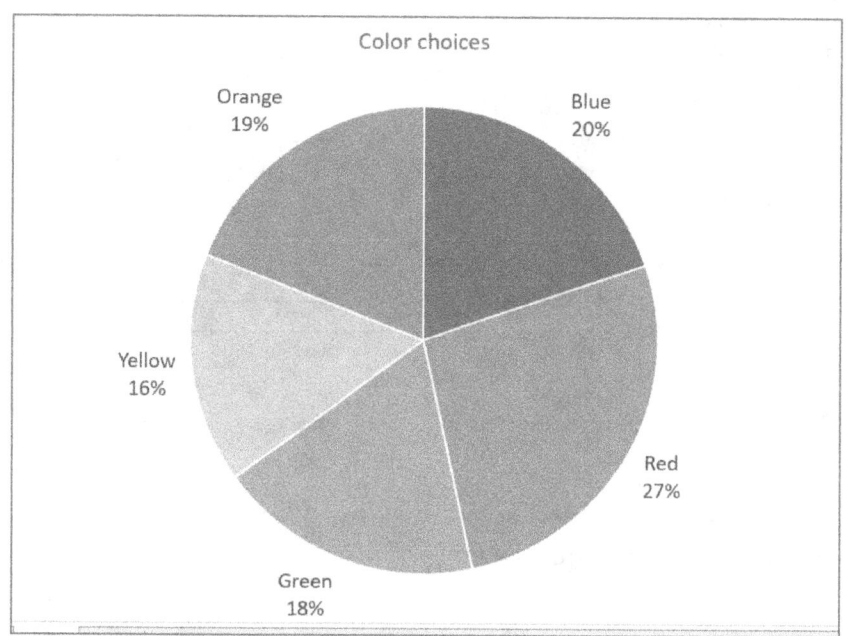

Figure 12: Excel's Pie Chart feature computes relative frequencies

Relative Frequencies

The pie chart shown in Figure 12 displays percentages for each color. These percentages are called relative frequencies and are the portion of each category (color) in relation to the whole. The relative frequency is the ratio of each color to the entire number of observations expressed as a percentage.

With the data used to create the pie chart, 42 of 157 people chose red as their favorite color in the list. The relative frequency is then, 42/157 or 26.75%.

Frequency Table		
Color choice	# choosing	Relative Frequency
Blue	31	19.75%
Green	29	18.47%
Orange	30	19.11%
Red	42	26.75%
Yellow	25	15.92%
Total	157	100.00%

Figure 13: A frequency table with relative frequencies

To add relative frequencies to a frequency table, divide the number of each category chosen by the total number of observations in the dataset. You can check your math by ensuring that all the relative frequencies total to 1 or 100%.

CUMULATIVE RELATIVE FREQUENCIES

Frequency distributions display the relative frequency for each category. When all the relative frequencies are added, they sum to 1 or 100%. Cumulative relative frequencies add each category's relative percentage to the category or categories above. The result is still 100%, but 100% is reached with the last category.

As the illustration below shows, blue and green represent 38.22% of the choices. The color choices of blue, green, orange, and red accounted for 84.08% of the choices.

Frequency Table			
Color Choice	# choosing	Relative Frequency	Cumulative Relative Frequency
Blue	31	19.75%	19.75%
Green	29	18.47%	38.22%
Orange	30	19.11%	57.32%
Red	42	26.75%	84.08%
Yellow	25	15.92%	100.00%
Total	157	100.00%	

Frequency Distributions for Quantitative Data

Briefly exam the data in Figure 14. Constructing frequency distributions for the variables gender, residence, and favorite color would be relatively simple tasks because there are a limited number of categories for each these variables. Counting the number of observations that fall into each category and then graphing the frequencies or placing them into a table would not be extremely time consuming and the resulting table or graph would be easy to interpret as we have seen with the favorite color frequency distribution.

ID	First	Last	Gender	Pay	Residence	Favorite color
5344	C	Alcantar	Female	$ 11.90	Nampa	Blue
5927	T	Allen	Female	$ 10.65	Boise	Red
5447	S	Allison	Female	$ 13.65	Caldwell	Green
5356	K	Alverson	Female	$ 12.20	Nampa	Yellow
4848	B	Amos	Female	$ 13.22	Meridian	Orange
5854	N	Anderson	Female	$ 16.15	Meridian	Green
5393	J	Anderson	Male	$ 12.25	Nampa	Yellow
5312	E	Anderson	Female	$ 11.30	Caldwell	Red
5239	N	Angel	Female	$ 15.36	Boise	Green
5885	M	Araiza	Male	$ 26.22	Caldwell	Blue
5886	M	Araiza	Female	$ 17.15	Caldwell	Red
5728	C	Arens	Female	$ 17.23	Boise	Green
5144	P	Armiger	Female	$ 16.48	Boise	Yellow
5733	S	Arnold	Female	$ 17.23	Meridian	Orange
5874	G	Au	Male	$ 25.78	Boise	Red
5558	D	Austin	Male	$ 15.05	Boise	Red
5549	J	Bair	Female	$ 15.30	Nampa	Blue
5986	D	Baker	Male	$ 27.10	Meridian	Blue

Figure 14: A large dataset with a quantitative variable

Constructing a frequency distribution table for the variable *pay*, because there are so many different values would not provide any useful information because the number of occurrences in each pay value would often be 1. You can see a small portion of a frequency distribution of *pay* values in Figure 15.

$	9.00	$9.50	$9.90	$9.95	$10.00	$10.05	$10.09	$10.10	$10.15	$10.20	$10.25	$10.30	$10.35	$10.40
Count of Pay	1	1	1	2	2	3	1	1	1	2	1	1	2	1

Figure 15: A frequency distribution of quantitative data

To create frequency distributions from quantitative data, especially when you are using continuous variables, it is better to create categories that contain a range of values, rather than counting each value. With the example data in the above illustration, you may decide to group the data by increments of $1. Then, you could count the number of observations from $9.00 to $9.99, $10.00 to $10.99 and so on.

Rather than guessing about the range and how many categories you should use for your values, there is a standard approach in statistics. That approach uses the following steps:

1. DETERMINE K, THE NUMBER OF CLASSES (CATEGORIES)

The classes are the ranges that will include the counts. Following the previous example with the "guessed" increments, there would be 17 classes with a range of $1 because the data ranges from

10.65 to 27.10. Again, rather than guessing, there a simple formula we can use to determine the number of classes in a dataset and that depends on the number of observations in the data.

The example dataset with the names, residences, and pay rates is available on the author's website in Excel format. It is called *Pay Survey*. It is not necessary to download the file, but you can use it to follow along if you like. The *Pay Survey* dataset contains 421 observations. The number of observations in the dataset is the key in determining how many classes to use in a frequency distribution of quantitative data.

The rule of thumb in determining the number of classes, k, is $2^k >= n$ (the number of observations). The process is shown in Figure 16 using the 421 observations in the example *Pay Survey* dataset.

Possible K	2K
2	4
3	8
4	16
5	32
6	64
7	128
8	256
9	512 >=421

Figure 16: Determining the number of classes

Using this rule, we determine the proper number of classes, or categories for this dataset is 9. Again, we computed k by increasing its value until 2^k is a value equal to or greater than the number of observations.

2. DETERMINE THE INTERVAL OR WIDTH

After you have determined the number of categories, k, you need to determine the width or range for each category. We used $1.00 as an example interval range. That may turn out to be correct, but there is there is a formula to determine the width or range.

To determine the range, you need to know the highest, (*maximum*), and lowest, (*minimum*), values in the dataset of the variable for which you are trying to create the frequency distribution. Unless you have downloaded the *Pay Survey* file, you will have to take my word that the highest value is $27.10 and the lowest value is $9.00. When you are working with a large dataset, Excel or other software program becomes very useful.

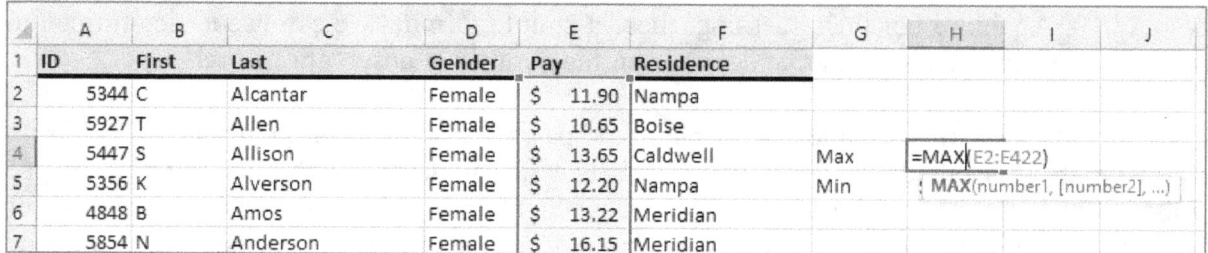

Figure 17: Determining the highest and lowest values for the variable

After we have determined the number of classes, k, and the highest and lowest values in the dataset, we can compute the proper width for each class. The formula we use is:

$$I \text{ (interval)} = (\text{Highest value} - \text{Lowest Value})/k$$

So for this dataset, $I = (27.10 - 9.00) / 9 = 2.01$.

Rounding this to an interval of 2 makes it simpler and easier to work with but doing this we will need to add a class. Otherwise we will miss the values greater than $27.00

The illustration in Figure 18 shows the frequency distribution with $k=10$ and $I = 2$ using Excel's pivot table feature. With a large dataset, counting records manually becomes very tedious.

Row Labels	Count of Pay
9-11	35
11-13	96
13-15	60
15-17	116
17-19	60
19-21	18
21-23	14
23-25	12
25-27	9
27-29	1
Grand Total	421

Figure 18: Completed Frequency Distribution for Quantitative data

Try it in Excel:

In Excel, the CountIf() function can also be used to create a frequency distribution.

ID	First	Last	Gender	Pay	Residence			
5344	C	Alcantar	Female	$ 11.90	Nampa			
5927	T	Allen	Female	$ 10.65	Boise			
5447	S	Allison	Female	$ 13.65	Caldwell	Max	$ 27.10	
5356	K	Alverson	Female	$ 12.20	Nampa	Min	$ 9.00	
4848	B	Amos	Female	$ 13.22	Meridian			
5854	N	Anderson	Female	$ 16.15	Meridian	9 - 11	35.00	
5393	J	Anderson	Male	$ 12.25	Nampa	11 - 13	=COUNTIF(E2:E422,">11")-COUNTIF(E2:E422,">13")	
5312	E	Anderson	Female	$ 11.30	Caldwell	13 - 15	COUNTIF(range, criteria)	
5239	N	Angel	Female	$ 15.36	Boise	15-17		
5885	M	Araiza	Male	$ 26.22	Caldwell			
5886	M	Araiza	Female	$ 17.15	Caldwell			

You can also use Excel's Pivot Table feature to quickly create frequency tables, even with numeric data.

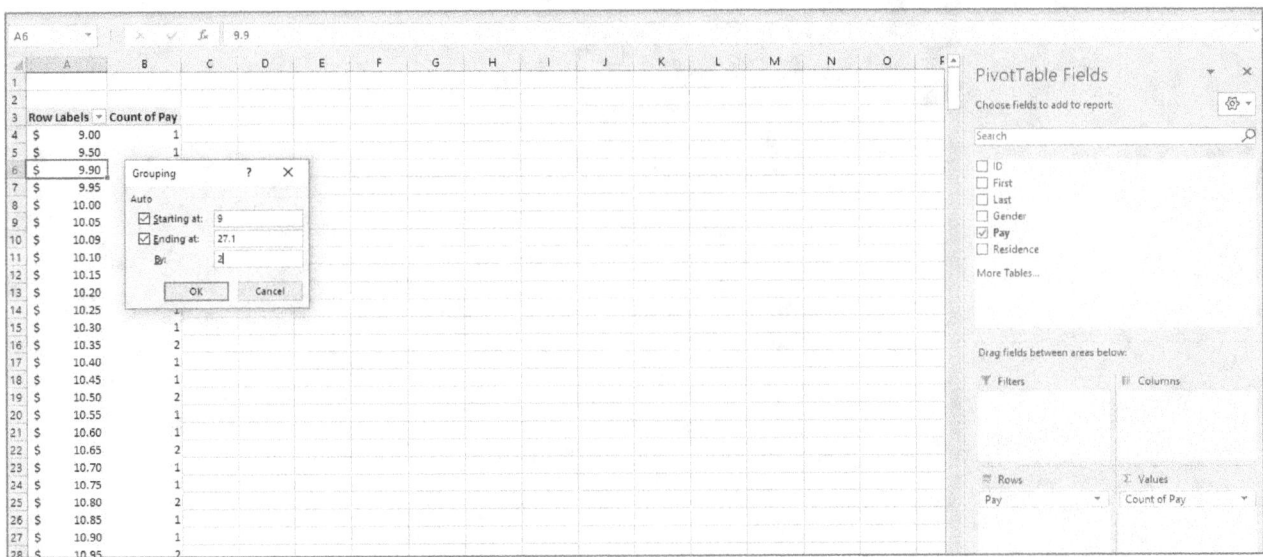

If you are familiar with Pivot tables, you may want to experiment with them for creating frequency distributions. If you are not familiar with them, there are many resources available, including several Excel workbooks I created to help you master Excel quickly and easily.

Chapter Highlights

Here are some of the important topics covered in this chapter:

- A frequency distribution counts the number of observations in each category.

- A frequency distribution can be displayed as a table, bar chart, column chart, or pie chart.

- Relative frequency is the number in the category divided by the total number of observations: 5/20, and is often displayed as a percentage, 25%.

- Frequency distributions for quantitative variables uses classes, k and ranges, i.

    o To determine the number of categories, k, take 2^k until the result is greater than or equal to the number of observations, n.

    o To determine the width or interval, find the difference between the highest and lowest values of the variable and divide the difference by the number of categories, k. *Interval = (Highest value – Lowest Value)/k*

Chapter 3: Descriptive Statistics – Numerical Methods

This chapter will cover:
- Measures of Central Tendency
  - Mean, Median, Mode
- Measures of Dispersion
  - Range, Variance, Standard Deviation
- Population vs. Sample Computations
- The Empirical Rule

Chapter 3: Descriptive statistics – Numerical Methods

Measures of Central Tendency

While frequency distributions tell us how many observations are in a category or within a specific range, there are additional methods we can use to describe the variables in our dataset. Among these are measures of central tendency. These measurements try to answer the question, "What does the average observation look like?" To answer this question, we will use the arithmetic mean, mode, and median.

THE ARITHMETIC MEAN

The mean is the "average" computation you learned to compute in your early school years. In statistics, we use "mean" because the word "average", to the "average" person (mathematicians excluded), can mean any one of the three computations we just mentioned.

In statistics, the mean is the mathematical center point of the dataset. You compute the mean by adding every value in the dataset and then dividing the sum by the number of items you added.

Population Mean - μ

$$\frac{\sum x}{N}$$

The population refers to all members of the group. (I know we already covered that, but repetition is essential in education.) The total number in the population is represented with a capital N. To compute the mean of the population, sum all the values of the variable of interest (x), and divide that by the population size, N. The population mean is represented by the Greek letter mu (μ).

Sample Mean – $\bar{x}$

$$\frac{\sum x}{n}$$

The number in the sample, the sample size, is represented with a lowercase n. We compute the sample mean the same as the population mean: sum all the values of the variable from the sample and divide the sum by the sample size, n. The sample mean is represented by the symbol $\bar{x}$.

As we progress through this subject, we will find that we need to know if we are looking at a sample mean, or a population mean. That is why we represent them differently, μ for population mean and $\bar{x}$ for sample mean. With some upcoming computations, we will compute the population and sample values differently.

Try it in Excel:

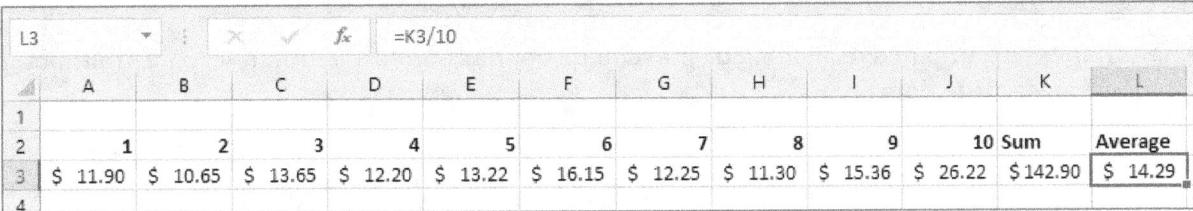

In this example, the 10 values for *x* are summed and then the sum is divided by *n*. (We are assuming this is a sample.)

In Excel, you can save a step by using the Average() function as shown. Note that Excel uses the term average, instead of mean, for this function.

It's Greek to me

To this point, we have not used too many Greek symbols, but we have introduced some symbols and letters used in statistics.

You probably wondered why we needed separate formulas for the population and sample means when they were the same formula. As you progress in statistics it will be very important to differentiate whether the values you are using in formulas are from a sample or a population. Some of the computations in this chapter and beyond are different if you are working with a population rather than a sample.

So, let's summarize the symbols we have used so far:

Understanding the symbols		
	Population	**Sample**
Number of observations	N	n
Mean	μ (mu)	$\bar{x}$ (x bar)

PARAMETERS VS. STATISTICS

Numerical characteristics of a population, such as the mean (μ), are called ***parameters***.

Numerical characteristics of a sample, such as the mean ($\bar{x}$), are called ***statistics***.

Do you remember the definitions of statistics from the first chapter? Now we can complete the second definition:

> **Statistics** are organized, analyzed, presented, or interpreted quantitative data (numbers) based on **sample** data.

So then,

> **Parameters** are organized, analyzed, presented, or interpreted quantitative data (numbers) based on **population** data.

The mean and outliers:

Because the mean includes every value in its computation, it is susceptible to outliers. To see how this works, assume that your statistics teacher has created a new quiz. There are seven students in your class, and they scored as follows:

Student	1	2	3	4	5	6	7
Score	100	68	68	68	67	67	66

After giving the quiz, your teacher asks for feedback on the difficulty of the quiz. Student #1, (probably you), has no problem with the quiz. It may have even been too easy. The other six students complain loudly. The quiz was far too difficult because none of them received a grade that will help them earn at least a "C" in the course, which is considered the lowest passing grade.

It's not my fault!

Your statistics teacher gets defensive. "I did some statistical analysis, and the mean quiz score was 72. That's exactly what the mean grade should be, a 'C'."

He is the teacher and you are the student, and if you are not student #1, you assume he knows what he is talking about. Still, something doesn't seem quite right with his analysis.

You have just experienced a limitation of the mean, which is commonly called the average. Because the mean includes every value in the dataset, it can be skewed higher or lower by a value or two that does not fit the rest of the values. In this example, student #1's score is not truly indicative of how the "average" student scored and neither is the computed mean of 72.

MEDIAN – LOCATIONAL CENTER POINT

Because the mean can be skewed by outliers, sometimes the median becomes a more accurate measure of the central tendency. The median is the value that lies in the middle location of the observations. There are an equal number of observations below and above the median.

To find the median:

1. Sort the values from low to high. (You will notice that I already did that with the test scores example).

2. Find the middle value. In the test scores example, there are 7 values. The median is a location #4, **three values higher** than #4 and **three values lower** than #4.

Student	1	2	3	4	5	6	7
Score	66	67	67	68	68	68	100

The median's value is 68. This is more representative of the actual scores than is the mean of 72. The median is not affected by outliers, while the mean is.

Finding the median with an even numbered dataset

Finding the center value is easy when there are an odd number of data points. As in the previous example with seven data points, the median is the value located at position #4. When you are working with an even number of data points, there is one additional step to finding the median. Here is how it works with the additional step.

1. Sort the values in ascending order, low to high.

2. Find the **location of the** middle value. In this example, the median is between location 3 and 4.

3. Compute the median by taking the mean (average) of two value in the middle position. In this case **(67 +68)/2 = 67.5.**

Student	1	2	3	4	5	6
Score	66	67	67	68	68	100

Try it in Excel:

If you have been following along in Excel, its Median function will do the math for you.

	A	B	C	D	E	F	G	H
1	Student	1	2	3	4	5	6	
2	Score	66	67	67	68	68	100	67.5

H2 fx =MEDIAN(B2:G2)

=median(B2:G2)

> **Jumping ahead**: In this example, the median is located at position 3.5. Knowing this, we can also compute the median's value by adding 50% (.5) of the difference between the value at position 3 and position 4 to the value of position 3. So, computing the median this way, we have 67 + ((68-67)*.5) = 67.5. Using the mean between the two values is an easier computation for now, but we will have to use this method in the next chapter.

MODE – THE MOST POPULAR VALUE

Another measure of central tendency is the *mode*. The mode is the value that occurs the most often or has the highest frequency of occurrences. In small datasets, you can look at the sorted data and see which value occurs the most often.

Our quiz results example poses an interesting problem because there is no one value that occurs the most often. In this example, it is a tie. The scores of 68 and 67 both occur twice. In this case, we would describe this data as bimodal (two modes).

Student	1	2	3	4	5	6
Score	66	67	67	68	68	100

Try it in Excel:

Excel's Mode() function works well, but only if there is only one mode in the dataset. If a dataset has two or more modes, Excel will return the one which it finds first. Excel's Array feature will allow you to find more than one mode, but that is a more advanced Excel feature, so we will not cover it in this book.

Measures of Dispersion

We are still discussing descriptive statistics. We started this chapter with measures of central tendency to describe the center point or average of the dataset. Now we are moving into measures of dispersion. Measures of dispersion still fall into the descriptive category and these measures tell us is how widely our data points vary, or how widely they are dispersed.

Why are measures of dispersion important? Here is another example to help you understand.

The town I live in is at the confluence of two rivers, the Clearwater and the Snake in Idaho. In the summer, a popular activity is driving up the Snake River for several miles and then floating down in a kayak or inner tube.

Assume for a moment you are visiting my town and someone has invited you to float down the Snake River. However, you cannot swim at all.

You express your concerns, but your host assures you with, "Oh, don't worry. The mean depth of the river is only four feet. You can get up and walk out at any time."

While your host correctly described the river with a descriptive statistic, the mean is not very helpful. What you really need to know is how deep the river is at its deepest point.

Either your host is trying to drown you, or they did not go very far in statistics class.

THE RANGE:

One of the most important, as easiest to compute, measures of dispersion is the range. It is the difference between the highest value and the lowest value in the dataset.

Student	1	2	3	4	5	6	7
Score	100	68	68	68	67	67	66

To compute the range of the student quiz data: 100 (highest score) – 66 (lowest score) = 34.

Try it in Excel:

Excel does not have a function to compute the range, but you can use the Max and Min functions as shown to easily compute the range of a dataset.

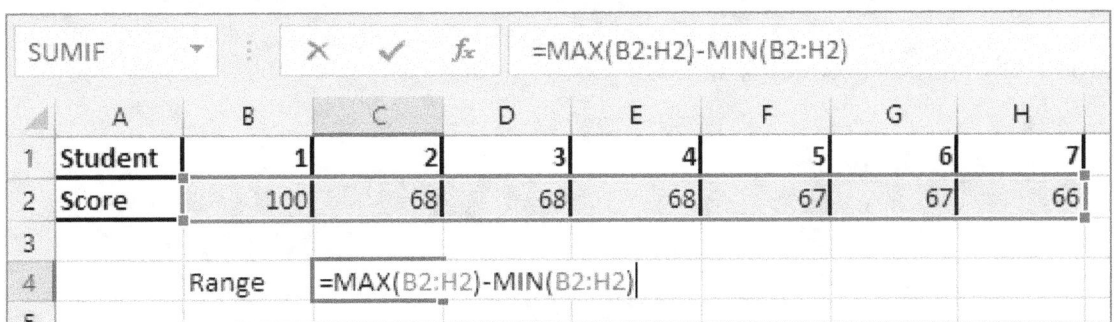

VARIANCE:

The variance is a measure of variability. It measures how widely the data varies by comparing each data point to the mean.

We use one of two formulas to compute the variance. We use one for computing the population variance and another for computing the variance of a sample.

Population variance:

$$\sigma^2 = \frac{\Sigma (x - \mu)^2}{N}$$

As you explore this formula, first notice the symbols it uses. We discussed μ and N as representing population values. We also designate the population variance with σ^2 (sigma squared).

To compute the variance, we take the difference between each value (*x*), and the population mean, μ. Then, we square each difference, sum up the squared differences, and then divide the sum by *N*, the number in the population.

To see how this works, we will assume the quiz scores we have been using represent a population. To compute the variance, we take the following steps:

1. Compute the population mean.
   (100+68+68+68+67+67+66)/7 = 72

   – we use μ to represent the mean because we are considering this a population, not a sample.

2. Compute the difference between each value and the mean:

x	μ	(x - μ)
100	72	28
68	72	-4
68	72	-4
68	72	-4
67	72	-5
67	72	-5
66	72	-6

3. Square the differences between x and μ.

Surviving Statistics

x	μ	(x - μ)	(x - μ)²
100	72	28	784
68	72	-4	16
68	72	-4	16
68	72	-4	16
67	72	-5	25
67	72	-5	25
66	72	-6	36

4. Sum the squared differences.

x	μ	(x - μ)	(x - μ)²
100	72	28	784
68	72	-4	16
68	72	-4	16
68	72	-4	16
67	72	-5	25
67	72	-5	25
66	72	-6	36
		Total	918

5. Divide the sum of the squared differences by the number in the population, N.

918 / 7 = 131.1429   so, the population variance is 131.1429.

The computed variance of 131.1429 may seem large considering the values we started with. Remember, this is the mean of the squared differences (deviations). Notice the deviation from the mean for the highest value, 100. It has a large deviation which results in a very large squared deviation. We discussed this value as an outlier that affected the mean. Outliers also cause the variance to be larger. But that is exactly why we compute the variance. The variance we computed tells us there is a large dispersion or variance in the values of these exam scores.

Sample variance:

$$s^2 = \frac{\Sigma (x - \bar{x})^2}{n - 1}$$

At first glance, the sample variance formula appears quite different than the population variance. However, when we remember that samples use different symbols, there is truly very little difference. Let's look at the same and different portions of this formula.

First, notice that sample variance is designated with s^2 rather than σ^2. The other symbol to notice is $\bar{x}$, which represents *sample* mean rather than *population* mean. So, the sample variance begins with the sum of the squared differences between each value and the mean, sample mean in this case. However, the real difference between the sample and population variance is the denominator.

Rather than dividing the sum of the squared differences by the number in the population, N, we divide by one less than the number in the sample, $n - 1$.

We will use the quiz scores to compute sample variance. We will assume the seven students represent a subset of a larger statistics class. To compute the variance of the exam scores, we take the following steps:

1. Compute the sample mean, $\bar{x}$.
   $(100+68+68+68+67+67+66)/7 = 72$
   – we use $\bar{x}$ to represent the mean because we are considering this a sample.

2. Compute the difference between each value and the sample mean, $\bar{x}$:

x	$\bar{x}$	(x - $\bar{x}$)
100	72	28
68	72	-4
68	72	-4
68	72	-4
67	72	-5
67	72	-5
66	72	-6

3. Square the differences between x and $\bar{x}$.

x	$\bar{x}$	(x - $\bar{x}$)	(x - $\bar{x}$)²
100	72	28	784
68	72	-4	16
68	72	-4	16
68	72	-4	16
67	72	-5	25
67	72	-5	25
66	72	-6	36

4. Sum the squared differences.

x	μ	(x - μ)	(x - μ)²
100	72	28	784
68	72	-4	16
68	72	-4	16
68	72	-4	16
67	72	-5	25
67	72	-5	25
66	72	-6	36
		Total	918

5. Divide the sum of the squared differences by one less than the number in the sample, $n - 1$.

   918 / 6 = 153   so, the sample variance is 153.

You may notice the sample variance is larger than the population variance. Of course, this makes sense when you remember you subtracted 1 from the sample size, making the denominator smaller. The official reason for subtracting 1 from the sample size is to reduce potential error (bias) from the values computed from samples, (statistics), and the population parameters.

Try it in Excel:

It is a good idea to understand how to compute the variance on your own, so you are familiar with how this computation works. Of course, Excel has built-in functions to compute both the sample and population variance.

Within Excel's Statistical function category, you will find the VAR.P and the VAR.S functions. The "P" of course stands for population and the "S" for sample.'

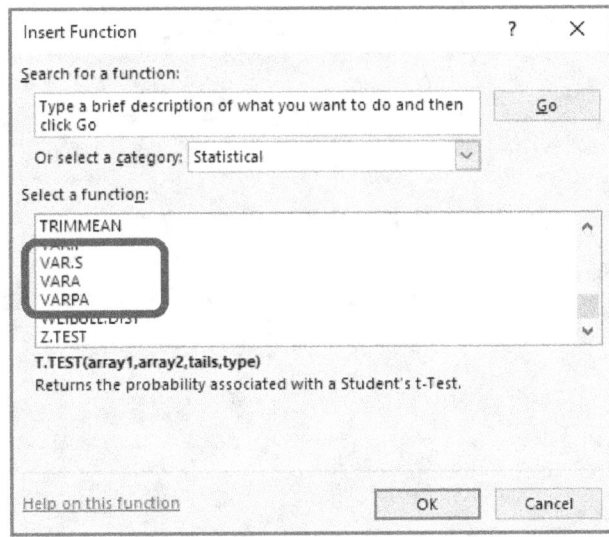

As you can see in the figure below, the values Excel computed matched our computations for both the population and sample variance.

	A	B	C	D	E	F	G	H
1	Student	1	2	3	4	5	6	7
2	Score	100	68	68	68	67	67	66
3								
4	Var.P	131.143						
5	Var.S	153						

B5: =VAR.S(B2:H2)

STANDARD DEVIATION:

In common statistical usage, the standard deviation is used more often than the variance. Once you have computed the variance, computing the standard deviation is very easy. The standard deviation is the square root of the variance.

Population Standard Deviation:

$$\sigma = \sqrt{\sigma^2}$$

$$\sigma = \sqrt{131.143} \quad = 11.4518$$

Sample Standard Deviation:

$$s = \sqrt{s^2} \quad = 12.3693$$

Or, as expressed by the entire formula for sample standard deviation:

$$s = \sqrt{\frac{\Sigma(x - \bar{x})^2}{n - 1}}$$

Try it in Excel:

Excel has built-in functions for population standard deviation and sample standard deviation. You can find these functions in the Statistical category.

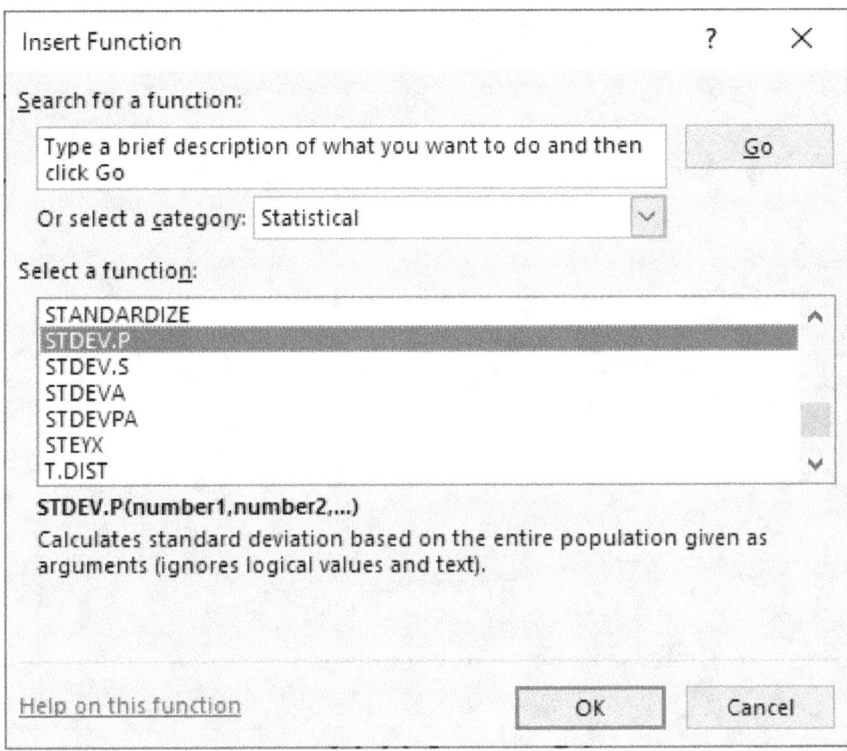

As you progress in statistics, you will find that you are using the standard deviation frequently. As you interpret data based on the standard deviation, you will find that a relatively small standard deviation indicates that most of the data points lie rather close to the mean.

THE EMPIRICAL RULE

You will spend considerable time in your statistics class dealing with data distributed within a normal, bell-shaped curve. In a normal curve, the highest point is the mean and, within a normal distribution, most of the observations will fall near the mean.

The empirical rule allows you to quickly "guestimate" how many of the observations fall within certain values. For example, using the illustration of the normal curve in Figure 19, the mean is 100 and the standard deviation is 10. The empirical rule tells us that 68% of the observations fall within + or – one standard deviation of the mean, or between 90 and 110 in this example.

The empirical rule also states that 95% of the observed values will fall within + or – two standard deviations, or with this example, between 80 and 120. Almost all, 99.7% of the observed values will fall between + or – three standard deviations of the mean, or between 70 and 130 here.

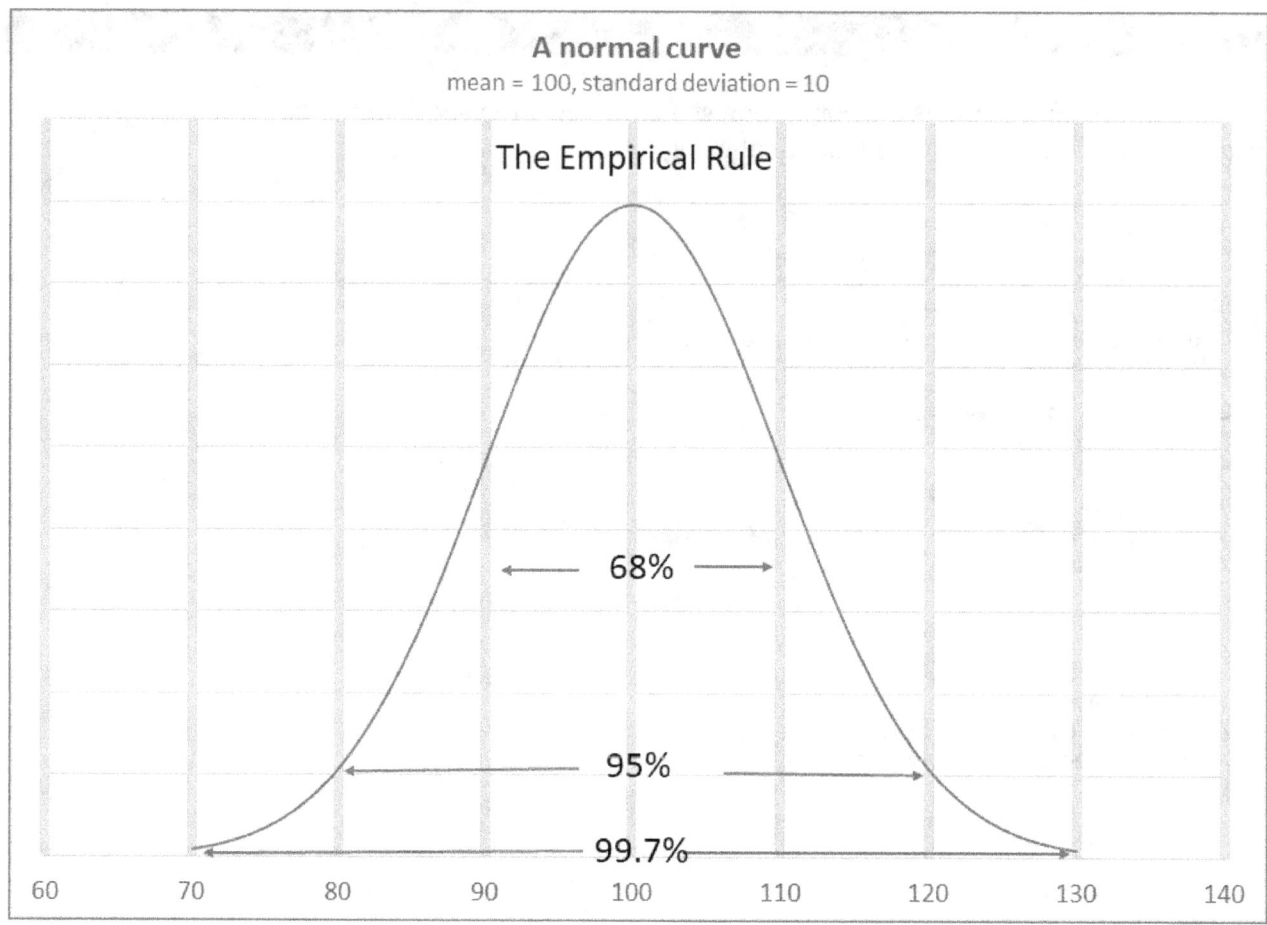

Figure 19: A normal curve illustrating the empirical rule

So what's the big deal?

Using the empirical rule, we can estimate values that will appear in data sampled from the population. In other words, knowing that 95% of the data points fall within + or − two standard deviations of the mean, we could sample someone in this population completely at random and be 95% sure their value was between 80 and 120.

Even though we are still working with descriptive statistics officially, you may be able to see how the empirical rule could help us begin to infer things about our data. If we know the mean and standard deviation, then we can guess a range of values and be accurate up to 68%, 95%, or 99.7%, depending on the level of accuracy we need and the number of standard deviations we are willing to extend beyond the mean.

Chapter Highlights

Here are some of the important things covered in this chapter:

- Measures of central tendency include:
    - Mean – the central point of the values numerically – average
    - Median – the locational center point of the values
    - Mode – value or values that occur in the data most frequently

- Parameters vs statistics:
    - Population values are called parameters
    - Sample values are called statistics
    - Different symbols are used to designate a value as a statistic or a parameter such as μ for population mean and $\bar{x}$ for sample mean

- Measures of dispersion tell us how widely the data varies:
    - Range: Largest value – Smallest value

    - Population variance:
    $$\sigma^2 = \frac{\Sigma (x - \mu)^2}{N}$$

    - Sample variance:
    $$s^2 = \frac{\Sigma (x - \bar{x})^2}{n - 1}$$

    - Population Standard Deviation: $\sigma = \sqrt{\sigma^2}$

    - Sample Standard Deviation: $s = \sqrt{s^2}$

- The empirical rule states that approximately:
    - 68% of the values lie within +/- 1 standard deviation from the mean
    - 95% of the values lie within +/- 2 standard deviations from the mean
    - 99.7% of the values lie within +/- 3 standard deviations from the mean

- Excel has built-in functions for mean (average), mode, median, standard deviation and variance.

Chapter 4: Exploring Data

This chapter will cover:
- **Measures of Position**
  - Location vs. Value
- **Skewness**
- **Scatterplots**

Chapter 4: Exploring Data

Measures of Position

We have already explored one measure of position, the median. The median, you should recall, is the middle point with 50% of the data above and 50% below the median. We can also refer to the median as the 50th percentile. We often describe data in terms of percentiles. Whether it is an SAT or other standardized test score, the height of a toddler, or an IQ measurement, the results are usually reported in percentiles. Reporting something as being in the 98th percentile means that only 2% of the items or individuals in that sample or population have higher values. Similarly, 98% of the members of the sample or population have lower values.

When working with measures of position you may hear the terms *quartile* and *decile* in addition to *percentile*. *Quartiles* measure the location in quarters, 25th, 50th, 75th, while *deciles* measure in 10th s. The median can be expressed as the 2nd quartile, or the 5th decile, and the 50th percentile.

The first step in finding the location of a percentile is to sort the data in ascending order, just as we did when we discussed the median.

Next, we find the location using this formula, where P is the percentile you want to locate.

$$L_p = (n+1)\frac{P}{100}$$

Position	1	2	3	4	5	6	7	8	9	10	11	12
Test score	50	68	71	74	76	79	83	85	91	92	98	100

Let's try some examples.

1. Find the location of the 25th percentile, the 1st quartile:

$$(12+1)\frac{25}{100} = 3.25$$

2. Find the location of the 90th percentile, the 9th decile:

$$(12+1)\frac{90}{100} = 11.7$$

LOCATION VS. VALUE

When working with measures of position, it is important to note that this is a two-step process. The formula to determine the *position* should not be confused with the *value* at that position, which you may also need to determine.

Let's try some examples.

1. Find the value associated with the 25th percentile, 1st quartile:

    First, we find the location. We did this earlier and calculated the location to be 3.25. So, the value of the 25th percentile is a number between the value at position 3 and the value at position 4.

    We compute the value of location 3.25 by multiplying the difference of the values at 3 and 4 by .25. Then, we add the result to the value at position 3. The value of the 25th percentile then is:

    $$71 + .25(74 - 71) = 71.75$$

    Our example dataset does not have an exact value at the 25th percentile, but this is the calculated value of the 25th percentile.

2. Find the location of the 90th percentile, 9th decile:

    We computed the location earlier with, (13) 90/100 = 11.7.

    This informs us that the value of the 90th percentile is the value at position 11, plus .70 times the difference between that value and the value at location 12.

    The value at location #11 is 98. The value at location #12 is 100. The value of the 90th percentile is:

    $$98 + .70(100 - 98) = 99.4$$

Skewness

Data does not always follow a symmetrical normal curve, such as the curve in the center of Figure 20. When the data is symmetric, the median and the mean are the same. The illustration on the left is *negatively skewed*, which means the mean is smaller than the median (negatively skewed). The curve on the right, is *positively skewed*, which indicates the mean is larger than the median.

 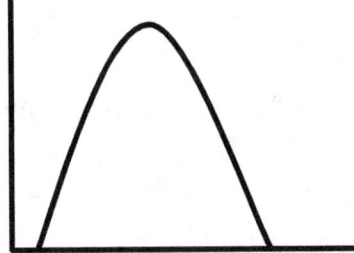 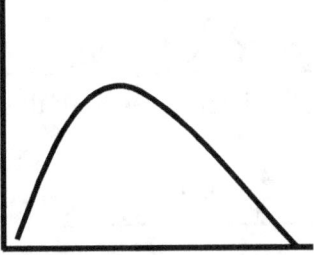

Figure 20: Examples of negatively skewed, symmetrical, and positively skewed curves

Drawing curves of our datasets could give us some indication that our data is skewed. Because determining the degree of skewness would be difficult to do just by looking at the curve, there are some formulas that allow us to numerically determine the direction and level of skewness in our data.

In chapter two, we introduced the *Pay Survey* data file with 421 observations. If you examine the descriptive statistics table created by Excel in Figure 21, you will see that that *mean* is slightly larger than the *median* in this dataset. This indicates that this data is positively skewed, but the difference is small, so we know it is at least slightly skewed. In the descriptive statistics, you will see it that Excel also computed a *skewness* value of .8965, a small positive number. Skewness values of less than -1 or greater than 1 indicate the data is highly skewed, either negatively or positively.

	A	B	C	D	E	F	G	H	I
1	ID	First	Last	Gender	Pay	Residence			
2	5344	C	Alcantar	Female	$ 11.90	Nampa			
3	5927	T	Allen	Female	$ 10.65	Boise		Descriptive Statistics	
4	5447	S	Allison	Female	$ 13.65	Caldwell			
5	5356	K	Alverson	Female	$ 12.20	Nampa		Mean	15.41853088
6	4848	B	Amos	Female	$ 13.22	Meridian		Standard Error	0.176330468
7	5854	N	Anderson	Female	$ 16.15	Meridian		Median	15.24
8	5393	J	Anderson	Male	$ 12.25	Nampa		Mode	17.23
9	5312	E	Anderson	Female	$ 11.30	Caldwell		Standard Deviation	3.617998705
10	5239	N	Angel	Female	$ 15.36	Boise		Sample Variance	13.08991463
11	5885	M	Araiza	Male	$ 26.22	Caldwell		Kurtosis	0.804386635
12	5886	M	Araiza	Female	$ 17.15	Caldwell		Skewness	0.896586238
13	5728	C	Arens	Female	$ 17.23	Boise		Range	18.1
14	5144	P	Armiger	Female	$ 16.48	Boise		Minimum	9
15	5733	S	Arnold	Female	$ 17.23	Meridian		Maximum	27.1
16	5874	G	Au	Male	$ 25.78	Boise		Sum	6491.2015
17	5558	D	Austin	Male	$ 15.05	Boise		Count	421
18	5549	J	Bair	Female	$ 15.30	Nampa			

Figure 21: Descriptive Statistics created by Excel

We will examine two different methods to compute skewness. Sometimes they will produce slightly different results. We will look at the Pearson's coefficient of skewness and the method software programs such as Excel use.

PEARSON'S COEFFICIENT OF SKEWNESS

Pearson's coefficient uses both the median and the standard deviation to compute skewness.

$$\frac{3(\bar{x} - Median)}{s}$$

Using the values in the Pay Survey data file, skewness is computed as:

$$\frac{3(15.42-15.24)}{3.618} = .14925$$

This value does not match the number Excel computed, but both skewness values indicate the data is slightly positively skewed.

SOFTWARE SKEWNESS COMPUTATION

$$\frac{n}{(n-1)(n-2)}\sum\left(\frac{x-\bar{x}}{s}\right)^3$$
This formula requires a few more computations than the Pearson's method, but is easy to use with software programs like Excel.

Try it in Excel:

You can access Excel's Skew() function in the Statistical function category or by simply typing the function.

	A	B	C	D	E	F	G
1	ID	First	Last	Gender	Pay	Residence	
2	5344	C	Alcantar	Female	$ 11.90	Nampa	
3	5927	T	Allen	Female	$ 10.65	Boise	0.89659
4	5447	S	Allison	Female	$ 13.65	Caldwell	
5	5356	K	Alverson	Female	$ 12.20	Nampa	
6	4848	B	Amos	Female	$ 13.22	Meridian	
7	5854	N	Anderson	Female	$ 16.15	Meridian	
8	5393	J	Anderson	Male	$ 12.25	Nampa	

(G3: =SKEW(E2:E422))

Descriptive Statistics	
Mean	15.41853
Standard Error	0.17633
Median	15.24
Mode	17.23
Standard Deviation	3.617999
Sample Variance	13.08991
Kurtosis	0.804387
Skewness	0.896586
Range	18.1
Minimum	9
Maximum	27.1
Sum	6491.202
Count	421

Excel also has a Descriptive Statistics summary function available with the Analysis Toolpak add-in. This add-in is included with Excel but must be enabled to use it.

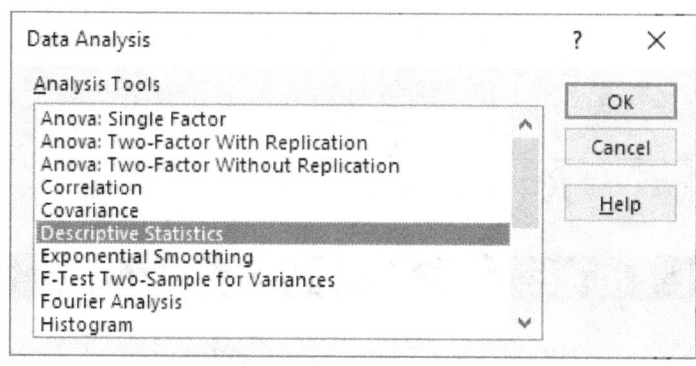

Relationships between Two Variables

Statistical analysis is often used to determine if two or more variables are related in any way. For example, if you sell hot dogs on a street corner, it would probably be safe to assume there is a relationship between the number of dogs you sell each day and the number of people that walk past your stand. Placing your hot dog stand on a street with no traffic would not be a smart business move.

A SCATTER PLOT

Lemonade Sales		
Day	Temperature	# glasses sold
Sunday	72	49
Monday	71	50
Tuesday	68	44
Wednesday	71	49
Thursday	73	55
Friday	67	42
Saturday	74	60
Sunday	70	49
Monday	71	50
Tuesday	68	44
Wednesday	71	49
Thursday	73	55
Friday	69	42
Saturday	74	60

One way to describe a relationship between two variables is to create a scatter diagram that plots both variables. Let's move from hot dogs to lemonade. Larry runs a lemonade stand. The table displays the daily high temperature and the number of glasses Larry sold for two weeks. It seems likely that a relationship exists between the daily high temperature and the amount of lemonade sold.

Scatter plots are useful tools to find correlations between variables. Although you can create a scatter plot on your own, tools like Excel make the process easier. Figure 22 shows a plot of the two variables, the number of glasses sold and the daily high temperature. The linear trend line, a feature of Excel, clearly indicate a relationship between these two variables.

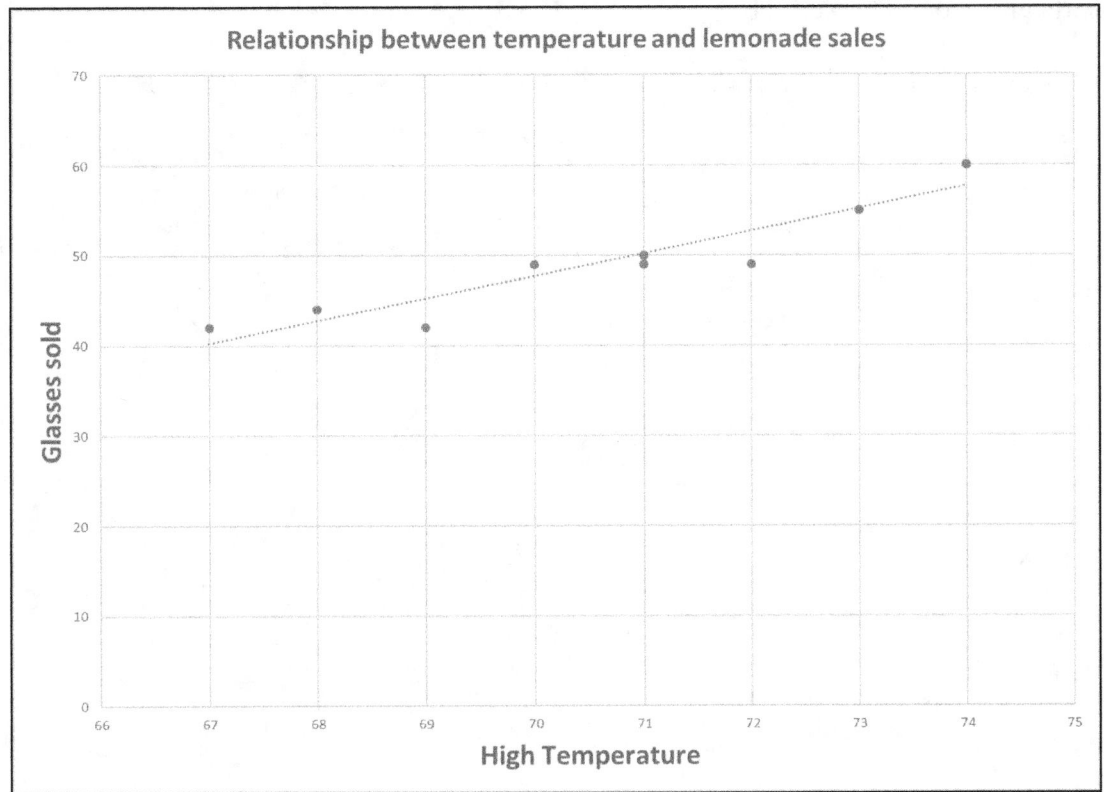

Figure 22: A Scatterplot

CORRELATION COEFFICIENT

The correlation coefficient is a value that numerically represents the relationship displayed in the scatter plot. The value expresses how strong the relationship or *correlation* is between two or more variables. The correlation coefficient, r, between the temperature and amount of lemonade Larry is 0.948. This indicates a very strong positive relationship between these two variables. We can clearly see this relationship in the scatter plot in Figure 22. You will learn how to compute a correlation coefficient in an upcoming chapter.

Try it in Excel:

You can easily create a scatterplot in Excel by selecting the two variables and then choosing Scatter from the Charts group on the Insert tab.

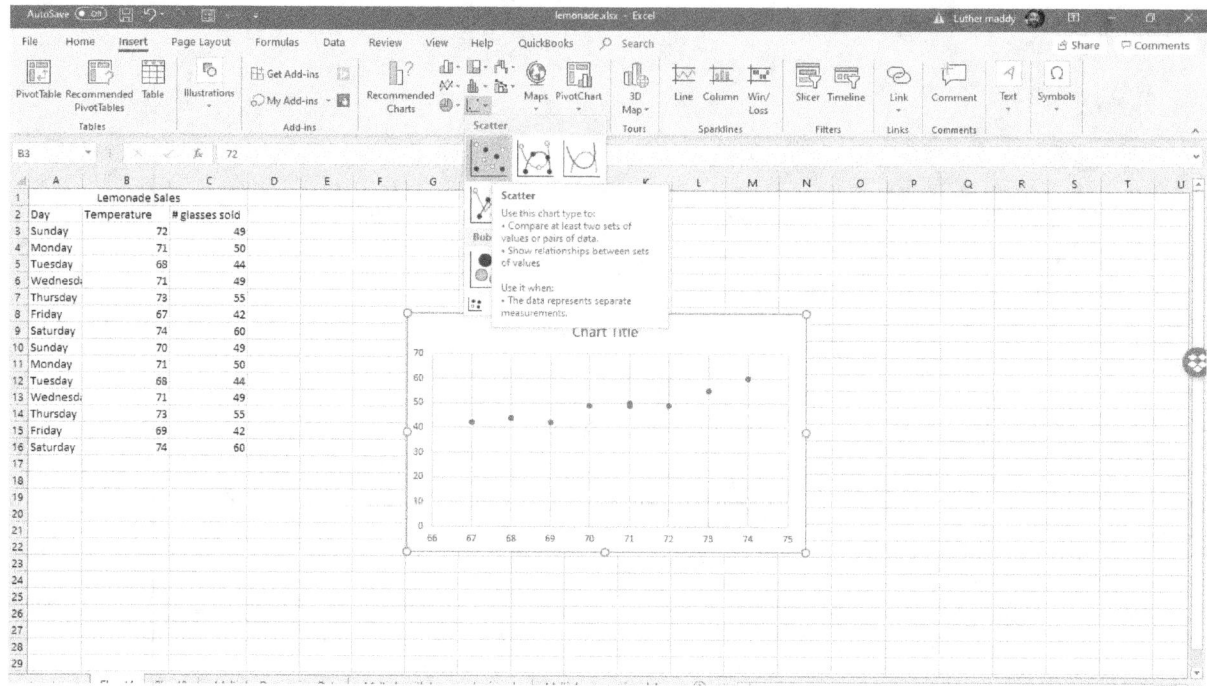

Chapter Highlights

- Measures of position can measure the locational point, or value of a percentile, decile, or quartile. The median is the 50th percentile.

    o The formula to find the location is: $$L_p = (n+1)\frac{P}{100}$$

    o To find the value associated with the location:
    1. First, we find the location with the above formula.
    2. If the position is not a whole number, take location 5.8 for example
        - Round down to the next whole number, 5.8 become 5.
        - Take the value at location 5 and add 80% of the difference between the values at location 5 and location 6.
            Value at 5 + .8(Value at 6 – Value at 5) = Value at location 5.8

- Skewness is a measure of how far the data deviates from a normal distribution (perfectly normal curve). It can be calculated with Pearson's coefficient of skewness or with the equation used by software programs like Excel.

    o Pearson's coefficient of skewness: $$\frac{3(\bar{x} - Median)}{s}$$

    o Software skewness computation: $$\frac{n}{(n-1)(n-2)}\sum\left(\frac{x-\bar{x}}{s}\right)^3$$

    - You can easily compute skewness in Excel with its Skew() function.

- A scatterplot is a visual representation of the relationship between two variables.

Surviving Statistics

Chapter 5: Exploring Probability Concepts

This chapter will cover:

Methods of Assigning Probability:
  Classical, Relative, Subjective

Rules of Addition

Rules of Multiplication

Counting Principles:
  Combinations & Permutations

Contingency Tables

Chapter 5: Exploring Probability Concepts

Probability is the likelihood of something happening - being struck by lightning for example. Probability ranges from 0 to 1 and is usually represented as a percentage from 0% to 100%. We often commonly express probability as the odds or chances of some occurrence.

Some examples of probably include:
- The odds of flipping a coin and having it land as tails.
- The likelihood that your favorite college football team wins the national championship.
- The chances of you drawing a king from a full deck of 52 cards.
- The chances of rolling a single die and it coming up six.
- Then odds of you winning the lottery.

Probability is a very important aspect of inferential statistics. Remember inferential statistics *infers* or makes educated guesses about populations based on samples.

> Inferential statistics is a tool businesses use to attempt to predict what percentage of people will buy a new product.

> Inferential statistics is a tool epidemiologists use to estimate what percentage of people will become infected with a new virus.

> Inferential statistics is a tool psychologists use to determine the percentage of people who will suffer from depression if they become unemployed.

Probability is based on experiments and outcomes. The experiment is the process that creates the outcome, tossing a coin for example. The outcome is the result, landing as tails.

Methods of Assigning Probability

There are three methods: classical, relative frequency (empirical), and subjective.

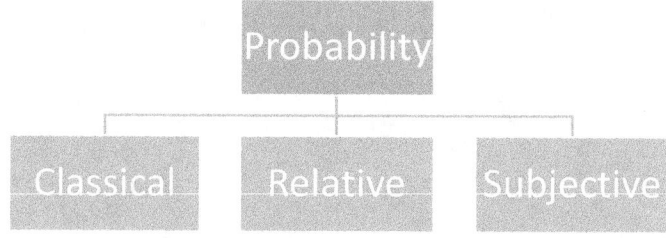

Figure 23: Three ways of assigning probability

CLASSICAL PROBABILITY

Classical probability is used when there are a set and finite number of outcomes and all outcomes are equally likely. For example, when you toss a coin, you will get either heads or tails. With classical probability, there is no guessing. You know the probability of the outcome before the experiment begins.

RELATIVE FREQUENCY OR EMPIRICAL PROBABILITY

Empirical probability determines probability based on repeated experiments or existing knowledge. For example, assume that 60 of last 100 customers in a coffee shop ordered a large coffee. What is the likelihood (probability) that the next customer will order a large coffee? Based on relative frequency, the probability is 60%.

SUBJECTIVE PROBABILITY

Subjective probability is based on intuition, guesses, and hunches, and has limited value in statistical research and analysis. Subjective probability begins with gathering all possible information and then, based on that information making a subjective guess about the odds of something occurring. For example, before the professional football season begins you gather information about your favorite team and the other teams and then subjectively choose what you believe to be that team's chances of winning the big game at the end of the season.

Check your understanding:

Choose the method used to assign probability in each of the following examples:

1. The odds of picking the queen of hearts from the top of a deck cards.

2. The chances of being struck by lightning in the United States this year.

3. At your first horse race, you like the looks of horse #4 and think the odds of it winning are good.

Answers:
1. We can compute the exact probability. There are 52 cards in a deck and one queen of hearts, so the probability is 1/52. This is an example of the classical method.

2. Every year approximately 270 people in the US are struck by lightning. With a population of approximately 330 million, your chances would be 270/330 million. This is an example of using relative frequency to determine probability.

3. You are an amateur and make a guess based on the appearance of the horse. This is an example of subjective probability.

Computing Probabilities

IMPORTANT TERMS

Before we begin discussing probability, here are some terms you should become familiar with.

Experiment
The process that generates the occurrence of one of several possible results.

Outcome
The result of an experiment. For example, the number you get from rolling a single die once.

Event
A collection of experiment outcomes. The values recorded from the experiment, rolling a die six times.

Collectively Exhaustive
When an experiment is conducted, at least one of the possible outcomes must occur. For example, when you roll a die, one number: 1, 2, 3, 4, 5, or 6 must come up.

Mutually Exclusive
One outcome result precludes any other possible outcome at the same time. For example, you cannot roll a 2 and a 4 on a die at the same time.

Independent Events
The occurrence of one outcome has no effect on the probability of another event occurring. For example, rolling a 1 on a die does not influence the value on the next or future rolls. After rolling a 1, you are just as likely to roll another 1 as you are to roll a 6.

Now that we have some of the terminology out of the way, let's dig into computing probabilities.

COMPLIMENT RULE

When assigning probabilities, the sum of all possible outcomes must equal, 1 or 100%. Knowing this, the compliment rule lets you assign probability for an outcome by knowing the probability of that outcome not occurring.

The compliment rule tells us the probability of an outcome is equal to 1 – the probability of not that outcome: $P(A) = 1 - P(\sim A)$. Note: we are using the $\sim$ symbol to represent "not".

For example, the probability of not drawing the queen of hearts off the top of a deck of cards is 51/52. So, the probably of drawing the queen of hearts is:

$$P(\text{Queen of Hearts}) = 1 - P(\sim \text{Queen of hearts}) \text{ or } 1 - 51/52 = 1/52$$

RULES OF ADDITION

Addition for Mutually Exclusive Events
When two events are mutually exclusive, to compute the probability of either one event or another, simply add the two probabilities. That is expressed as:

P(A or B) = P(A) + P(B)

What is the probability of rolling a single die and getting either a 1 or a 3?

$$P(1) = 1/6 \qquad P(3) = 1/6$$

$$P(1 \text{ or } 3) = 1/6 + 1/6 = 1/3$$

This is a mutually exclusive event, because you cannot roll a 1 and a 3 at the same time. These two possibilities do not intersect.

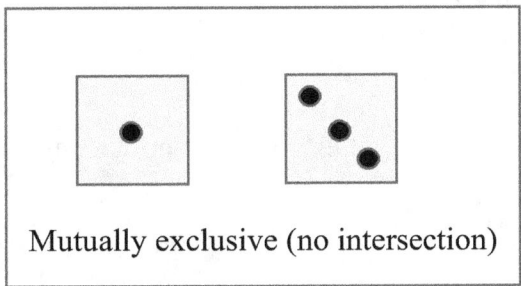
Mutually exclusive (no intersection)

Addition for Non-Mutually Exclusive Events
One or more non-mutually exclusive events can occur at the same time. The occurrence of one does not preclude the other from occurring, it is possible these events may overlap.

For example, assume you conducted a survey of all the residents in your city (*population*) to see if they owned cats, dogs, or neither. Your survey found that 25% of those in the survey own dogs and 20% own cats. However, 5% of those you surveyed own both a dog and a cat. This is shown in the illustration below.

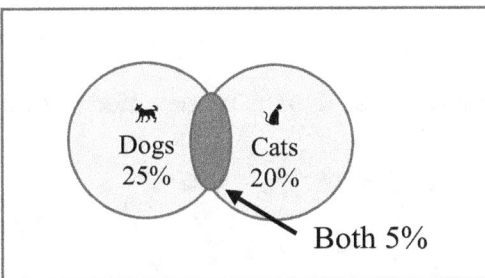

What is the probability that a resident of your city has either a cat or a dog?

$$P(A \text{ or } B) = P(A) + P(B) - P(A \text{ and } B)$$

$$P(\text{Cat or Dog}) = P(\text{Cat}) + P(\text{Dog}) - P(\text{Cat and Dog}) = .25 + .20 - .05 = .40$$

So, the likelihood of any resident in your city owning a cat or a dog is 40%.

RULES OF MULTIPLICATION

Multiplication for Independent Events

When events are independent, one outcome has no effect on other outcomes, the probability of two events occurring are the probability of the first event multiplied by the probability of the second.

$$P(A \text{ and } B) = P(A) * P(B)$$

For example, what is the probability of tossing a coin and having it come up heads both times?

$$P(A\text{-heads and } B\text{-heads}) = .5 * .5 = .25$$

If the events are independent, the same concept applies no matter how many events are included.

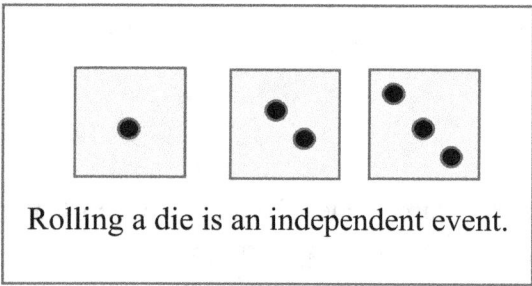

Rolling a die is an independent event.

For example, what is the probability of rolling one die three times and having it come up 1, 2, and 3 in that order.

$$P(1, 2, 3) = P(1) * P(2) * P(3) = 1/6 * 1/6 * 1/6 = 0.00463$$

Multiplication for Dependent Events

Events are dependent when one outcome does affect other outcomes. This is represented with the *given that* symbol |. So, for dependent events:

P(A and B) = (P(A)) (P(B | A))

The probability of both events occurring is the probability of the first one, A, multiplied by the probability of B occurring, given that A has already occurred.

Surviving Statistics

As an example, consider a bowl that contains 20 pieces of candy. Ten are blue and ten are yellow.

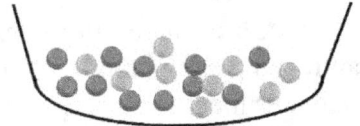

What is the probability of choosing two candies at random and having them both be blue?

$$P(Blue_1 \text{ and } Blue_2) = P(Blue_1) * P(Blue_2 | Blue_1) = 10/20 * 9/19 = 0.23684$$

Randomly choosing a blue candy the second time, given that the first candy was also blue is 9/19 because there are only 9 blue candies left and 19 total left.

The second blue outcome is an example of **conditional probability**. Conditional probability is the probability of a specific event occurring given that another event, $Blue_1$ in this case, has already occurred.

Contingency Tables

When we attempt to classify observations with two or more categories, we often create contingency tables. Contingency tables lay out the responses in a crosstab format. Computing probabilities in contingency tables use the dependent events formula.

As an example, assume you want to see if male students have different fast-food preferences than female students. In your survey, you ask the students to select their favorite fast-food restaurant from a group of three. The results are then classified by male/female and then counts are displayed for each favorite restaurant. The results are displayed in the contingency table below.

Gender	McRonald's	Burger Queen	Chicken Deluxe	Total
Male	15	7	23	45
Female	9	**18**	36	**63**
Total	24	25	59	**108**

The contingency table clearly displays dependent events. For example, the value of 23 in the Chicken Deluxe column is dependent on the gender being male. Here are some examples of using the table to compute probabilities.

What is the probability of finding respondent who is female and prefers Burger Queen?

We again use this formula: $P(A \text{ and } B) = (PA)(P(B | A))$

So, $P(\textit{Female} \text{ and } \textit{Burger Queen}) = (P(\textit{Female}))(P(\textit{Burger Queen} | \textit{Female}))$
(63/108) (18/63) = 0.1667

Counting Principles

When we are conducting an experiment, it is often important to know how many possible outcomes can result. For example, if you toss a coin, there are two possible outcomes. Assume though that the experiment is tossing three coins, or four. (Not tossing one coin four times.) Then the number of possible outcomes becomes much larger. In determining the number of outcomes, three methods commonly used.

MULTIPLICATION FORMULA

When an experiment has multiple steps, or a certain way of doing one thing and another way or ways of doing another, you use the multiplication formula to determine the total number of outcomes. For example, think of an experiment of rolling two dice. The first die has 6 possible outcomes and so does the second.

Using the multiplication formula, the total number of outcomes is:

$$(\text{Outcome}_1)(\text{Outcome}_2) = (6)(6) = 36$$

As another example, assume your statistics teacher gives you a test with three questions and each has four choices. How many ways can you complete the test?

$$(3 \text{ questions})(4 \text{ answers}) = 12$$

COMBINATION FORMULA

The combination formula allows you to compute how many ways r items can be selected from a larger group, n. Assume you own seven movies and have decided to watch two in a row on a boring weekend. How many different combinations of those two movies are possible?

To answer this question, you would use the combination formula:

$$_nC_r = \frac{n!}{r!(n-r)!}$$

In this formula, ! means factorial. So, solving your movie binging dilemma looks like:

$$_7C_2 = \frac{7!}{2!(7-2)!} \quad \text{or} \quad _7C_2 = \frac{(7)(6)(5)(4)(3)(2)(1)}{((2)(1))((5)(4)(3)(2)(1))} = 21$$

So, selecting two movies from the seven you own results in 21 possible movie combinations.

Surviving Statistics

PERMUTATION FORMULA

When you wanted to know how many combinations were possible in watching two out of the seven movies you own, using the combination formula yielded twenty-one possible combinations. With the combination formula, the order in which you watched to two movies was not important.

However, if instead of only asking what combination of two movies you could watch, you also wanted the combinations to include a specific order of the movies watched. In other words, you want to create a list that tells you to watch one specific movie first and then watch another specific movie second. In that case, you would use the permutation formula.

The permutation formula computes how many combinations of r are possible from n when the order is important.

Using the same example, how many combinations of two movies from the seven you own are possible when you also want the order watching one first and then another?

The permutation formula:

$$_nP_r = \frac{n!}{(n-r)!} = 42$$

So, when the order is specified as this movie first and that on second, there are 42 possible combinations.

Check your understanding:

Ponder the following questions:

1. Why, when you are working with the same two movies out of the seven you own, are there more permutations than combinations?

2. You locked your bicycle with a 4-digit combination lock with digits 0 – 9 but forgot how to open it. How many different four-digit patterns are possible?

3. You have two pairs of shoes, three pairs of pants, and five shirts. How many different outfits can you create from your wardrobe?

Surviving Statistics

Answers:

1. When the order is important, you use permutations. This means that a pair of two movies can be chosen more than once with permutations. For example, you could watch movie A first and then movie B. Or, you could watch B first and then A. So, the paring A – B, and B – A are possible using permutations.

    With the combination formula, the order is not important. So, selecting A and then B, or B and then A represent only one combination when the order is not a consideration.

2. Since the order is important and numbers do not repeat, you will compute this with the permutation formula. With n = 10 and r = 4, the number of permutations possible is 5040.

3. This is an example of when to use the multiplication formula for independent events. The number of outfits is: 2 * 3 * 5 = 30

Try it in Excel:

Excel has functions for computing with Combinations or Permutations.

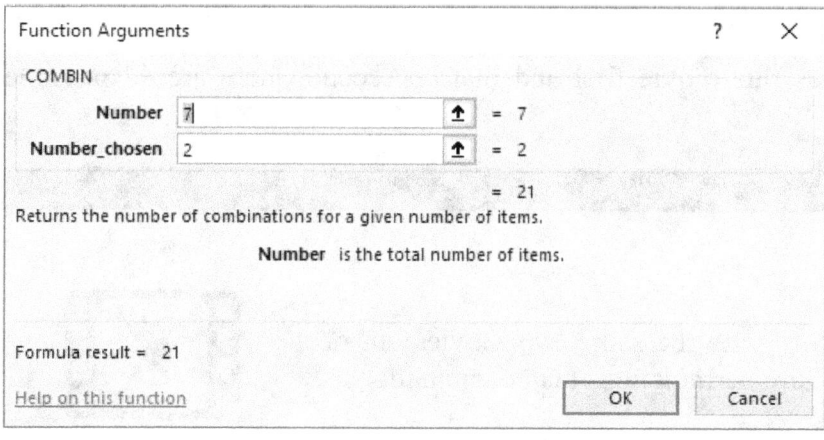

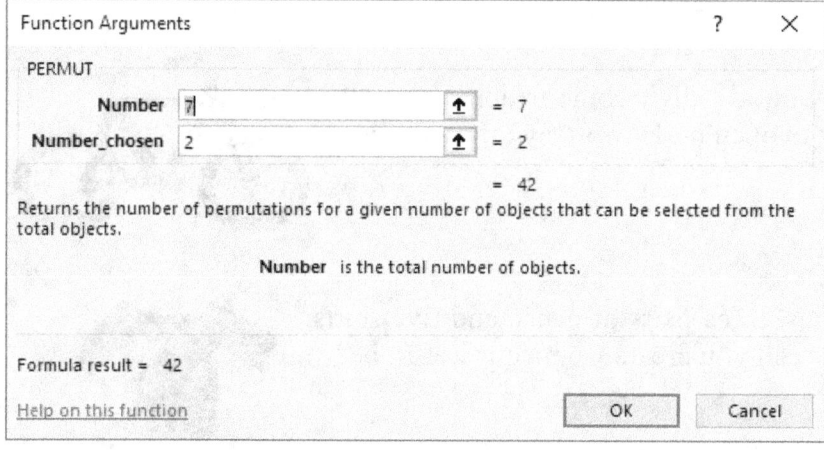

Chapter Highlights

Probability is the chances of something happening. Probability varies from 0 to 1, or 0% to 100%.

- **Addition for Mutually Exclusive Events**

When two events are mutually exclusive, to compute the probability of either one event or another, you add the two probabilities. The formula is: $P(A \text{ or } B) = P(A) + P(B)$

- **Addition for Non-Mutually Exclusive Events**

One or more non-mutually exclusive events can occur at the same time. The occurrence of one does not preclude the other from occurring.
The formula is: $P(A \text{ or } B) = P(A) + P(B) - P(A \text{ and } B)$

- **Multiplication for Independent Events**

When events are independent, one outcome has no effect on other outcomes. The probability of two events occurring are the probability of the first, times the probability of the second.
$P(A \text{ and } B) = P(A) * P(B)$

- **Multiplication for Dependent events**

Events are dependent when one outcome does affect other outcomes. This is represented with the given symbol, |. So, for dependent events:
$P(A \text{ and } B) = (PA) * (P(B | A))$

- **Counting Principles**
    - **Multiplication Formula**

        When an experiment has multiple steps, or a certain way of doing one thing and another way or ways of doing another, you use the multiplication formula to determine the total number of outcomes. Using the multiplication formula, the total number of outcomes is: (Outcome$_1$ * Outcome$_2$)

    - **Combination Formula**

        The combination formula allows you to compute how many ways r items can be selected from a larger group, n. The combination formula: $_nC_r = \dfrac{n!}{r!(n-r)!}$

    - **Permutation Formula**

        The permutation formula computes how many combinations of r are possible from n when the order is important. The permutation formula: $_nP_r = \dfrac{n!}{(n-r)!}$

Chapter 6: Discrete Probability Distributions

This chapter will cover:

Discrete Probability Distributions
Mean, Variance

Binomial Probability Distribution

Poisson Probability Distribution

Chapter 6: Discrete Probability Distributions

Probability Distributions

A probability distribution lists all possible outcomes from an experiment. It also lists the probability of each possible outcome. To be a true probability distribution the sum of all the probabilities listed must equal 1 or 100%.

Experiment: Toss a coin

Outcome	Probability
Heads	0.5
Tails	0.5
Total	**1**

Before we dive deeper into probability distributions, here a few terms we need to discuss.

RANDOM VARIABLES

A random variable is a numeric value that results by chance from an experiment. The result of rolling a die or flipping a coin is a random variable. The weight of a two-year-old, selected by chance, is also a random variable.

Discrete Random Variable

We previously defined a discrete variable as the result of counts. A discrete random variable is the result of an experiment that is selected by chance and the resulting value can be counted. An example of a discrete random variable could be the number of cars driving through an intersection in a ten-minute period.

Continuous Random Variable

A continuous variable can assume any value and often results from measurements. For example, the weight of a two-year old selected at random would be a continuous random variable. We will discuss probability distributions for continuous random variables in an upcoming chapter.

DISCRETE PROBABILITY DISTRIBUTIONS

A discrete random variable often assumes a value that can be counted. This is the case with rolling dice. To understand this concept of a discrete probability distribution, let's use the experiment of tossing 2 six-sided dice. The results are counted.

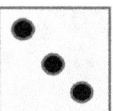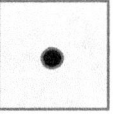

You cannot roll two dice and get 2.3 or 2.4. You will get only whole numbers, which is often the case of discrete variables. Two dice with six outcomes results in 36 possible outcomes, (6)(6) using the multiplication formula.

1. The first step in creating a probability distribution is to list all the possible outcomes and the number of ways or expected frequency of each outcome as shown.

Probability distribution: Roll two dice											
Outcome	2	3	4	5	6	7	8	9	10	11	12	Total
# of ways	1	2	3	4	5	6	5	4	3	2	1	36

2. The next step is to compute the probability of achieving each outcome. You do this by dividing the frequency of the outcome by the total number of outcomes, 2/36 for example. If all the probabilities sum to 1, you have accounted for all outcomes and have a probability distribution.

Outcome	2	3	4	5	6	7	8	9	10	11	12	Sum
# of ways	1	2	3	4	5	6	5	4	3	2	1	36
Probability	0.028	0.056	0.083	0.111	.139	0.167	0.139	0.111	0.083	0.056	0.028	1.00

Expected Value or Mean of a Discrete Probability Distribution

As you can see from the probability distribution, some outcomes are more likely to result from throwing two dice than others. The outcome with the highest probability is 7. If we toss two dice, what value do we expect, on average, to see?

$$\mu = \sum(xP(x))$$

To compute the mean or expected value of a discrete probability distribution, multiply each outcome by its probability and sum the results.

Outcome	# of ways	Probability	x P(x)
2	1	0.028	0.056
3	2	0.056	0.167
4	3	0.083	0.333
5	4	0.111	0.556
6	5	0.139	0.833
7	6	0.167	1.167
8	5	0.139	1.111
9	4	0.111	1.000
10	3	0.083	0.833
11	2	0.056	0.611
12	1	0.028	0.333
Sum	36	1.000	7

The expected value, or mean, of rolling two dice is 7.

Surviving Statistics

Variance of a Discrete Probability Distribution

The variance is a measure of dispersion as you learned previously. However, the formula you used in the previous chapter does not apply when dealing with a discrete probability distribution. The formula for computing the variance is:

$$\sigma^2 = \sum [(x - \mu)^2 P(x)]$$

The steps in computing this are:
1. Subtract the mean from the value of each outcome and square the difference.
2. Multiply the squared differences by the probability of that outcome.
3. Sum the squared differences multiplied by the probability.

Outcome	# of ways	Probability	x P(x)	(x - u)²	(x - u)² P(x)
2	1	0.028	0.056	25.000	0.694
3	2	0.056	0.167	16.000	0.889
4	3	0.083	0.333	9.000	0.750
5	4	0.111	0.556	4.000	0.444
6	5	0.139	0.833	1.000	0.139
7	6	0.167	1.167	0.000	0.000
8	5	0.139	1.111	1.000	0.139
9	4	0.111	1.000	4.000	0.444
10	3	0.083	0.833	9.000	0.750
11	2	0.056	0.611	16.000	0.889
12	1	0.028	0.333	25.000	0.694
Sum	36	1	7.000		5.833

The variance of this discrete probability distribution is 5.833.

The standard deviation is the square root of the variance, in this case: 2.415.

Try it in Excel:

Excel has no built-in functions to compute the mean and variance of a discrete probability distribution. However, it is a great calculator if you create the formulas manually. The image below shows one way to approach these calculations. The image displays the formulas used in those computations.

	A	B	C	D	E	F
1	Outcome	# of ways	Probability	x P(x)	(x - u)²	(x - u)² P(x)
2	2	1	=B2/$B$13	=C2*A2	=(A2-$D$13)^2	=E2*C2
3	3	2	=B3/$B$13	=C3*A3	=(A3-$D$13)^2	=E3*C3
4	4	3	=B4/$B$13	=C4*A4	=(A4-$D$13)^2	=E4*C4
5	5	4	=B5/$B$13	=C5*A5	=(A5-$D$13)^2	=E5*C5
6	6	5	=B6/$B$13	=C6*A6	=(A6-$D$13)^2	=E6*C6
7	7	6	=B7/$B$13	=C7*A7	=(A7-$D$13)^2	=E7*C7
8	8	5	=B8/$B$13	=C8*A8	=(A8-$D$13)^2	=E8*C8
9	9	4	=B9/$B$13	=C9*A9	=(A9-$D$13)^2	=E9*C9
10	10	3	=B10/$B$13	=C10*A10	=(A10-$D$13)^2	=E10*C10
11	11	2	=B11/$B$13	=C11*A11	=(A11-$D$13)^2	=E11*C11
12	12	1	=B12/$B$13	=C12*A12	=(A12-$D$13)^2	=E12*C12
13	Sum	=SUM(B2:B12)	=B13/$B$13	=SUM(D2:D12)		=SUM(F2:F12)

BINOMIAL PROBABILITY DISTRIBUTIONS

Binomial probability distributions are another type of discrete (countable) probability distribution. However, in a binomial distribution there are only two possible outcomes. A binomial distribution output might be pass or fail, yes or no, late or not late, purchase or not purchase, or even live or die. With a binomial probability distribution, the probability of any one success or failure is the same, no matter how many trials are done.

For example, Juan is a tropical fish breeder. He ships his fish from Florida to anywhere in the United States. From past experience, Joe has discovered that each fish he ships has a 95% chance of making it through the shipping process and arriving alive. Juan wants happy customers, so he throws in an extra fish or two with every order. This way he ensures the customer will receive at least the number of fish he or she ordered alive. However, Juan still needs to make a profit, so he wants to know exactly how many extra fish to ship.

So, if a customer orders 6 fish, what is the probability all fish will arrive alive and that none of the fish, 0, will be dead on arrival (DOA)?

What is the probability that 1 of the 6 fish shipped will die in the process?

What are the probabilities that 2, 3, 4, 5, or all 6 will die in transit?

To solve this dilemma, Juan resorts to something he learned about in his statistics course, the binomial probability formula.

The Binomial probability formula is:

$$P(x) = {_nC_x}\, p^x (1-p)^{n-x}$$

Where: p^x = the probability of x number of successes
$_nC_x$ = Combination, where *n* is the number of trials
p = the probability of a success on each trial

For Juan's example, we will consider a success (outcome we are concerned about) as arriving deceased. The probability of any one fish not making it is 0.05. So, what is the probability that none, 0, of the 6 fish will die in shipment?

Let's step through this computation.
We are calculating the probability of 0 fish dying during shipment, *P(x)* with *x* = 0

We are shipping 6 fish, so *n* = 6.

The probability of any one fish dying during shipment, *p*, is 0.05.

Any value to the power of 0 equals 1, so $p^0 = 1$.

The combination of *n* = 6 and *x* = 0 also equals 1.

$$P(0) = {_6C_0}\, p^0 (1-p)^{6-0} \quad = (1)(1)(1-0.05)5^5 = 0.73509$$

The probability that all the fish will live, none of the fish will die in shipment is 74%.

What is the probability exactly 1 fish will die in shipment?

$$P(1) = {_6C_1}\, p^1 (1-p)^{6-1} \quad = (6)(.05)(1-0.05)^6 = 0.2321$$

The probability that exactly 1 fish in 6 will not survive shipment is 23.21%.

What is the probability that the shipment of 6 fish will arrive with 2 DOA fish?
That computes to .0305.

Binomial Probability Distribution Tables
Binomial probability computations can be somewhat time consuming. Because of that, statistics textbooks often include tables like the one on the one shown in Figure 24. The binomial probability distribution tables are set up so that if you know the number of trials, six with Juan's fish example, and the probability of success, 0.05 in the example, you can then use the table to find the probability.

Examine the table in Figure 24. To solve the first scenario, 0 fish die in shipment, we first find the table a trial size, n of 6. This is the correct table. Then, locate the 0.05 probability column, the probability of any one success that Juan already knew. According to the table, the probability of 0 fish dying in shipment is 0.735, just what we computed.

When you use a Binomial table, first note that you are using the correct table based on the number of trials. Then, find the value where the number of successes and the probability intersect to find the probability of that number of successes.

x	0.05	0.1	0.2	0.3	0.4	0.5	0.6	0.7	0.8	0.9	0.95
0	0.735	0.531	0.262	0.118	0.047	0.016	0.004	0.001	0.000	0.000	0.000
1	0.232	0.354	0.393	0.303	0.187	0.094	0.037	0.010	0.002	0.000	0.000
2	0.031	0.098	0.246	0.324	0.311	0.234	0.138	0.060	0.015	0.001	0.000
3	0.002	0.015	0.082	0.185	0.276	0.313	0.276	0.185	0.082	0.015	0.002
4	0.000	0.001	0.015	0.060	0.138	0.234	0.311	0.324	0.246	0.098	0.031
5	0.000	0.000	0.002	0.010	0.037	0.094	0.187	0.303	0.393	0.354	0.232
6	0.000	0.000	0.000	0.001	0.004	0.016	0.047	0.118	0.262	0.531	0.735

n = 6

Figure 24: Binomial Distribution Table

To finish up with Juan, with a shipment of six fish, he will throw in two extra fish. By doing this, he is nearly 95% sure that his customers will receive all the fish they ordered, and possibly even an extra one or two.

Mean - Binomial Distribution
Computing the mean of a binomial distribution is not difficult. It is simply the number of trials, 6 for Juan's fish, multiplied by the probability of success for each trial, 0.05 in the example.

$$\mu = np$$

So, with Juan and his fish, the mean of fish expected to die in shipment is (6)(.05) = .3

Variance - Binomial Distribution
Like the mean, computing the variance of a binomial distribution is also easy. The formula to do so is:

$$\sigma^2 = np(1 - p)$$

So, for Joe the variance for fish arriving dead is:
(6)(.05)(1-.05) = .285

Surviving Statistics

Try it in Excel:

Excel's Binom.Dist() function will compute probabilities for x number of successes. You will need to provide the number of trials and the probability of success. As you see in the illustration, you can also use this function to create your own Binomial Probability Distribution tables. The "FALSE" at the end of the function informs Excel that you do not want to create a cumulative probability table, but instead want it to display the probability for each number of successes.

	D	E	F
			=BINOM.DIST(D4,$F$1,$F$2,FALSE)
1	# of trials		6
2	Probability of DOA		0.05
3	Successes	Probability	
4	0	=BINOM.DIST(D4,$F$1,$F$2,FALSE)	
5	1	0.2321	
6	2	0.0305	
7	3	0.0021	
8	4	0.0001	
9	5	0.0000	
10	6	0.0000	

The cumulative probability adds the probabilities for each number of successes. So, as you can see in this illustration, there is a 96.72% probability of having <=1 successes, which for our example, is fish dying in shipment.

	D	E	F
			=BINOM.DIST(D4,$F$1,$F$2,TRUE)
1	# of trials		6
2	Probability of DOA		0.05
3	Successes	Probability	Cumulative
4	0	0.7351	0.7351
5	1	0.2321	0.9672
6	2	0.0305	0.9978
7	3	0.0021	0.9999
8	4	0.0001	1.0000
9	5	0.0000	1.0000
10	6	0.0000	1.0000

POISSON PROBABILITY DISTRIBUTIONS

The Poisson distribution is considered a discrete distribution because it also relies on counting. The Poisson distribution also needs a specific interval of time or space. For example, the Poisson distribution is helpful to determine how many customers will come into a store in a specific period of time. The Poisson distribution could also be used to predict the number of potholes on a given stretch of freeway, or the number of bears in a square mile of forest wilderness.

The binomial distribution required that we have probability of success to determine the probability of exactly x successes. With the Poisson distribution, the probability is determined with the mean number of occurrences. Once the mean is determined, the probability of x successes can be determined using the Poisson distribution formula, which is:

$$P(x) = \frac{\mu^x e^{-\mu}}{x!}$$

e is a constant: 2.71828.

To see how the Poisson distribution works, assume Kathy is considering hiring another barista for her drive-through coffee stand. It takes approximately five minutes to process an order from start to finish which means her current barista can adequately handle three customers every fifteen minutes. Kathy knows that her customers become impatient if they must wait too long. Yet, she wants to maximize her profits and does not want to hire another barista if unnecessary.

One of Kathy's busiest times is between 10:15 and 10:30 in the morning. To see if she needs to hire another barista, Kathy records the number of customers arriving each day during this time period for ten days.

Between 10:15 - 10:30 am	
Day	# of customers
1	2
2	4
3	1
4	1
5	2
6	7
7	5
8	4
9	3
10	1
Mean	3

Surviving Statistics

Using this information, Kathy sees that on average, three customers arrive during this fifteen-minute period. At first glance, it seems her current staffing of one barista is adequate. But, to be sure, Kathy wants to know the probability of four customers arriving during that time, which means one customer may end up choosing to purchase coffee elsewhere.

Using a mean of three, Kathy can now use the Poisson distribution to compute the probability of four customers arriving from 10:15 – 10:30.

$$P(x) = \frac{\mu^x e^{-\mu}}{x!} = ((3)^4(2.71828)^{-3}) / 4! = .16803$$

Kathy learns there is nearly a 17% chance of having exactly 4 customers show during this time period.

However, Kathy wants to know more. What is the probably of having 5 or 6 or 7 or even more customers show up during this time period?

Fortunately, like the Binomial distribution, most textbooks have Poisson distribution tables that allow you to easily look up a specific or cumulative probability. Using a table, such as the one illustrated in Figure 25, Kathy can get the vital information she needs.

mean = 3	
X	Probability
0	0.0498
1	0.1494
2	0.2240
3	0.2240
4	0.1680
5	0.1008
6	0.0504
7	0.0216
8	0.0081
10	0.0008
11	0.0002
12	0.0001
13	0.0000

Figure 25: Poisson distribution table with mean = 3

The probability of more than 3 customers arriving during this time period is easily calculated by (1 – the probability of 0, 1, 2, or 3). The cumulative probability of 0, 1, 2, or 3 customers is .6472. So, the probability of having more than 3 customers is 1 - .6472, or .3538.

Kathy decides the probability of losing customers is too great, more than 35%, so she decided to hire another barista in the morning. She will continue to track the number of customers to see if the mean number of customers she originally computed is accurate.

Try it in Excel:

Excel's Poisson.Dist() function allows you to compute an individual probability or create a Poisson distribution table. By changing the cumulative argument from False to True, you can also create a cumulative probability distribution.

	H	I	J	K
1	Mean	3		
2	X	Probability	Cumulative	
3		0	0.0498	0.0498
4		1	0.1494	0.1991
5		2	=POISSON.DIST(H5,$I$1,FALSE)	
6		3	0.2240	0.6472
7		4	0.1680	0.8153
8		5	0.1008	0.9161
9		6	0.0504	0.9665
10		7	0.0216	0.9881
11		8	0.0081	0.9962
12		10	0.0008	0.9970
13		11	0.0002	0.9972
14		12	0.0001	0.9973

Formula bar: =POISSON.DIST(H5,$I$1,FALSE)
Tooltip: POISSON.DIST(x, mean, cumulative)

Chapter Highlights

- **Probability Distribution**

A probability distribution lists all possible outcomes from an experiment and the probability of that outcome.

- **Random Variables**

A random variable is a numeric value that results, by chance, from an experiment.

  o **Discrete Random Variable**

   A discrete random variable is the result of an experiment that is selected by chance and the value can be counted.

  o **Continuous Random Variable**

   A continuous variable can assume any value and often results from measurements.

- **Discrete Probability Distributions**

Every possible outcome from an experiment that results in discrete random variables, such as rolling a six-sided die.

  o **Expected Value or Mean of a Discrete Probability Distribution**

   To compute the mean or expected value of a discrete probability distribution, you multiply each outcome by its probability and sum the results.

   $$\mu = \sum(xP(x))$$

  o **Variance of a Discrete Probability Distribution**

   The formula for computing the variance of a discrete probability distribution is:

   $$\sigma^2 = \sum[(x - \mu)^2 P(x)]$$

- **Binomial Probability Distributions**

With binomial probability distributions there are only two possible outcomes.

The Binomial probability formula is: $P(x) = {_nC_x}\, p^x (1-p)^{n-x}$

   Where
   $P(x)$ = the probability of x number of successes
   ${_nC_x}$ = Combination
   p = the probability of a success on each trial

You can also use readily available tables to lookup Binomial probabilities.

- **Mean - Binomial Distribution**

    The mean of a binomial distribution is the number of trials multiplied by the probability of success for each trial.

    $$\mu = np$$

- **Variance - Binomial Distribution**

    The formula to compute the variance of a binomial distribution is:

    $$\sigma^2 = np(1 - p)$$

• **Poisson Probability Distributions**

The Poisson distribution relies on a specific interval of time, space, or area. With the Poisson distribution, the probability is determined with the mean number of occurrences. Once the mean is determined, the probability of x successes can be determined using the Poisson distribution formula which is:

$$P(x) = \frac{\mu^x e^{-\mu}}{x!}$$

e is a constant: 2.71828

Chapter 7: Continuous Probability Distributions

This chapter will cover:
- Uniform Probability Distributions
- Normal Probability Distributions
- Standard Normal Distribution (Z Scores)

Chapter 7: Continuous Probability Distributions

Uniform Distributions

In a uniform probability distribution, the probability of any outcome is equal to that of any other. To illustrate this, assume you have arrived at a bus stop. The bus is not there and so you must wait for it to return. The schedule of times is missing, but you know a bus arrives at this stop every 60 minutes. You do not know when the bus was last here, so your waiting time can be from 0 to 60 minutes. And, the probability of your waiting time, with your lack of knowledge, is uniform for any of the times in that range.

Uniform probability distributions are represented as a rectangle with the minimum and maximum values.

Since this is a uniform distribution, the probability of any point in the rectangle is the same as any other. The probability of you waiting 10 minutes is equal to your waiting 50 minutes.

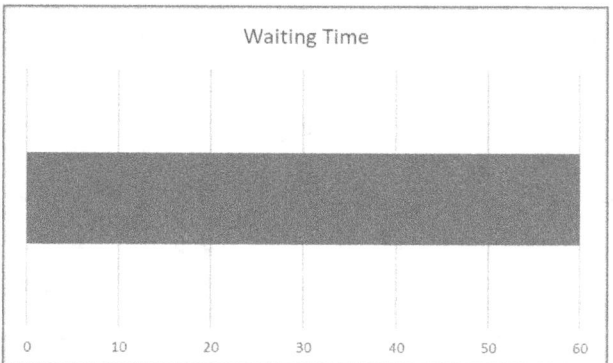

As you are waiting for the bus, you will invariably want to know the probabilities of different waiting time scenarios.

For example:
1. What is the mean (expected) wait time?

2. What is the probability you will have to wait 5 minutes or less?

3. What is the probability you will have to wait longer than 45 minutes?

Here are the computations to answer your questions.

The mean of a uniform distribution:

$$\mu = \frac{a+b}{2}$$ - where a is the minimum and b is the maximum value

Your expected wait time is: 0 + 60 / 2 = 30

To answer the other questions, take the following steps:

1. Compute the probability of any one point in the distribution. The probability is computed as:

$$P(x) = \frac{1}{b-a}$$

The probability of any specific wait time is 1/60.

2. Determine the number of units you are determining the probability for.
     5 minutes or less: 0 + 5 = 5.

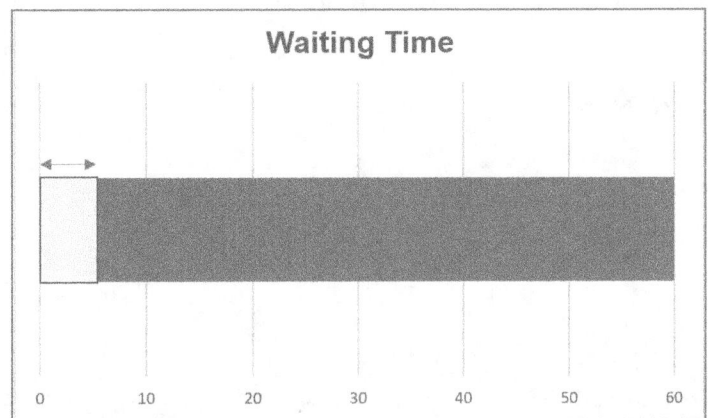

3. Multiply the number of units by the probability of each unit, minutes in this case.
     (5)(1/60) = 5/60 = .08333

What is the probability you will have to wait longer than 45 minutes?

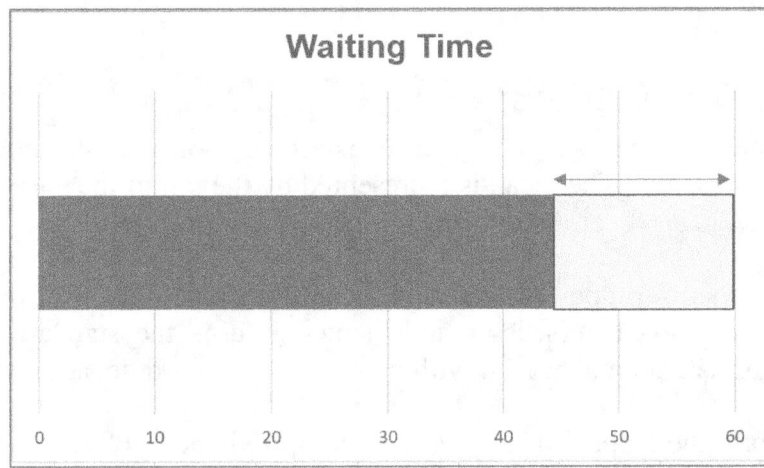

a. 60 − 45 = 15
b. (15)(1/60) = 15/60 = .25

What is the probability you will have to wait 30 minutes or less?
   a. 0 + 30 = 30
   b. (30)(1/60) = 30/60 = .5

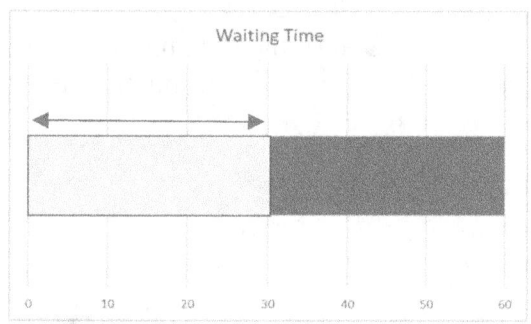

What is the probability you will have to wait between 20 and 30 minutes?
   a. 30 – 20 = 10
   b. (10)(1/60) = 10/60 = .1667

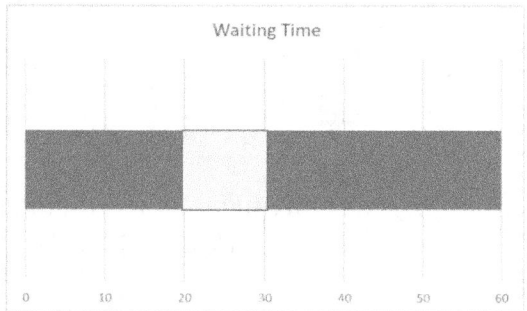

Normal Probability Distributions

A normal curve is represented by a bell-shaped curve. In a perfectly normal distribution, the mean, mode, and median are all the same. The mean is represented by the top of the curve and most of the data points are distributed closely around the mean.

The standard deviation is also an important value in a normal probability distribution. The empirical rule, which we previously discussed in Chapter 3, uses the standard deviation to determine the percentage of data points that fall within 1, 2, or 3 standard deviations of the mean.

Just to refresh your memory, the empirical rule states that approximately 68% of the data points fall within + or – 1 standard deviation of the mean. Using the illustration below, with the empirical rule in mind, if we pick a data point in this distribution at random, there is a 68% probability that the value will fall between 90 and 110.

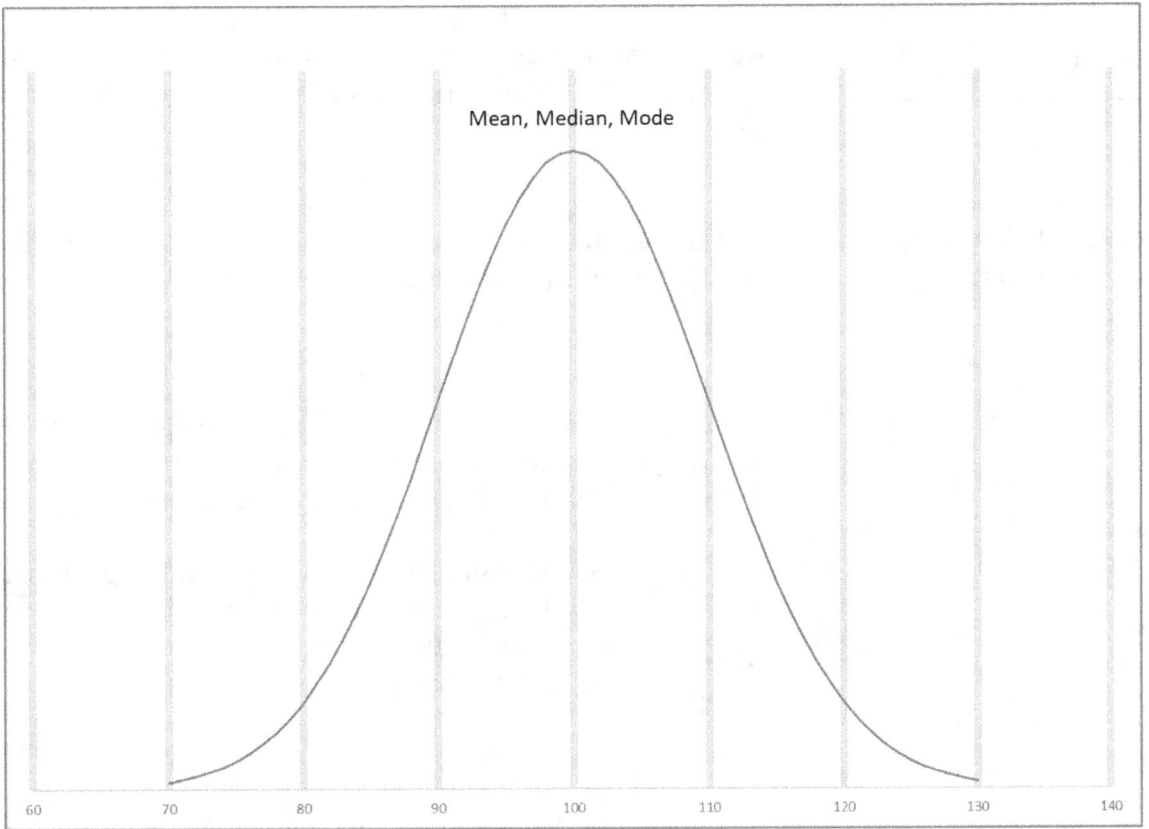

Figure 26: A normal distribution curve

STANDARD NORMAL DISTRIBUTION (Z SCORES)

The standard normal distribution and z scores makes it easier to compute probabilities and areas under the curve. Instead of working with innumerable values for means and standard deviation, the computed z score reports how many standard deviations a value varies from the mean. Knowing the z score for a value, we can then compute the area under the curve, or probability.

This will make more sense as we step through an example.
The formula for computing a z score is:

$$z = \frac{x - \mu}{\sigma}$$

Using the previous illustration, the mean for this distribution is 100 and the standard deviation is 10. Please note that we used σ to designate the standard deviation, which indicates this is a population value, not one derived from a sample.

Assume we want to know the probability of choosing a data point higher than 110 at random. With this illustration, it is easy to determine that the value of 110 is 1 standard deviation higher than the mean. With "real world" examples, it isn't always that easy so let's step through computing a score with those values and then build on that example.

$$z = \frac{110 - 100}{10} = 1$$

Recall that the z score tells us how many standard deviations the value varies from the mean, which in this case is 1.

The illustration below shows the area of the normal curve we are trying to quantify, the area (probability) of having a z score > 1, which represents a value of 110.

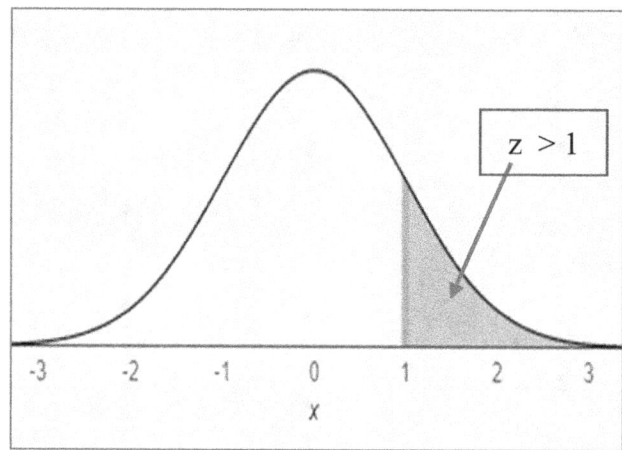

The shaded area is all the area to the right of the z score because we are concerned with a score of 110 or higher, which is a z score of 1.

To determine the probability of finding a score of 110 or greater in this distribution, we need to know how much of the curve is taken up with the shaded area.

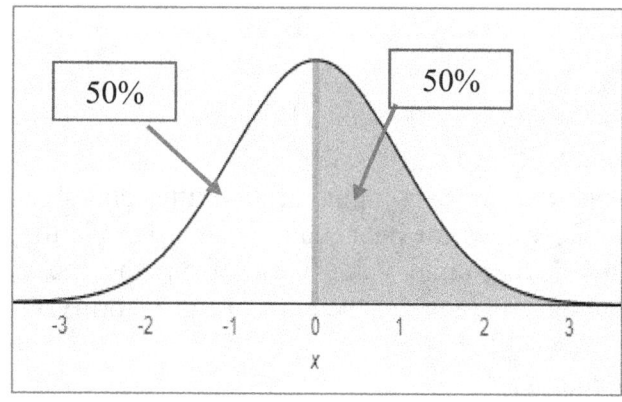

To determine the area within the curve for a z score, the first important concept is to realize that the curve is essentially divided by 2. The mean marks the position where half of the curve represents a value less than the mean and the other half represents values greater than the mean.

It should be obvious that a z score of 1.0 will represent less than 50% of the curve's area.

Computing the probability, area of the curve, for a z score manually is somewhat complex and time consuming. Using software like Excel makes this easier.

Most textbooks include a table similar to Figure 27. You can easily use this table to find probabilities for z scores. We will use this table as we continue to follow the example.

Area under the normal curve										
z	0.00	0.01	0.02	0.03	0.04	0.05	0.06	0.07	0.08	0.09
0	0.0000	0.0040	0.0080	0.0120	0.0160	0.0199	0.0239	0.0279	0.0319	0.0359
0.1	0.0398	0.0438	0.0478	0.0517	0.0557	0.0596	0.0636	0.0675	0.0714	0.0753
0.2	0.0793	0.0832	0.0871	0.0910	0.0948	0.0987	0.1026	0.1064	0.1103	0.1141
0.3	0.1179	0.1217	0.1255	0.1293	0.1331	0.1368	0.1406	0.1443	0.1480	0.1517
0.4	0.1554	0.1591	0.1628	0.1664	0.1700	0.1736	0.1772	0.1808	0.1844	0.1879
0.5	0.1915	0.1950	0.1985	0.2019	0.2054	0.2088	0.2123	0.2157	0.2190	0.2224
0.6	0.2257	0.2291	0.2324	0.2357	0.2389	0.2422	0.2454	0.2486	0.2517	0.2549
0.7	0.2580	0.2611	0.2642	0.2673	0.2704	0.2734	0.2764	0.2794	0.2823	0.2852
0.8	0.2881	0.2910	0.2939	0.2967	0.2995	0.3023	0.3051	0.3078	0.3106	0.3133
0.9	0.3159	0.3186	0.3212	0.3238	0.3264	0.3289	0.3315	0.3340	0.3365	0.3389
1	0.3413	0.3438	0.3461	0.3485	0.3508	0.3531	0.3554	0.3577	0.3599	0.3621
1.1	0.3643	0.3665	0.3686	0.3708	0.3729	0.3749	0.3770	0.3790	0.3810	0.3830
1.2	0.3849	0.3869	0.3888	0.3907	0.3925	0.3944	0.3962	0.3980	0.3997	0.4015
1.3	0.4032	0.4049	0.4066	0.4082	0.4099	0.4115	0.4131	0.4147	0.4162	0.4177
1.4	0.4192	0.4207	0.4222	0.4236	0.4251	0.4265	0.4279	0.4292	0.4306	0.4319
1.5	0.4332	0.4345	0.4357	0.4370	0.4382	0.4394	0.4406	0.4418	0.4429	0.4441
1.6	0.4452	0.4463	0.4474	0.4484	0.4495	0.4505	0.4515	0.4525	0.4535	0.4545
1.7	0.4554	0.4564	0.4573	0.4582	0.4591	0.4599	0.4608	0.4616	0.4625	0.4633
1.8	0.4641	0.4649	0.4656	0.4664	0.4671	0.4678	0.4686	0.4693	0.4699	0.4706
1.9	0.4713	0.4719	0.4726	0.4732	0.4738	0.4744	0.4750	0.4756	0.4761	0.4767
2	0.4772	0.4778	0.4783	0.4788	0.4793	0.4798	0.4803	0.4808	0.4812	0.4817
2.1	0.4821	0.4826	0.4830	0.4834	0.4838	0.4842	0.4846	0.4850	0.4854	0.4857
2.2	0.4861	0.4864	0.4868	0.4871	0.4875	0.4878	0.4881	0.4884	0.4887	0.4890
2.3	0.4893	0.4896	0.4898	0.4901	0.4904	0.4906	0.4909	0.4911	0.4913	0.4916
2.4	0.4918	0.4920	0.4922	0.4925	0.4927	0.4929	0.4931	0.4932	0.4934	0.4936
2.5	0.4938	0.4940	0.4941	0.4943	0.4945	0.4946	0.4948	0.4949	0.4951	0.4952
2.6	0.4953	0.4955	0.4956	0.4957	0.4959	0.4960	0.4961	0.4962	0.4963	0.4964
2.7	0.4965	0.4966	0.4967	0.4968	0.4969	0.4970	0.4971	0.4972	0.4973	0.4974
2.8	0.4974	0.4975	0.4976	0.4977	0.4977	0.4978	0.4979	0.4979	0.4980	0.4981
2.9	0.4981	0.4982	0.4982	0.4983	0.4984	0.4984	0.4985	0.4985	0.4986	0.4986
3	0.4987	0.4987	0.4987	0.4988	0.4988	0.4989	0.4989	0.4989	0.4990	0.4990

Figure 27: A normal distribution table

Using the table, the area under the curve associated with a z score of 1.0 is .3413.

After finding this value, you may be tempted to answer the question, "What is the probability of finding a score above 110?" with 34%. However, the answer is not quite that simple.

The area you located, .3413, represents the area of the curve from the mean to that z score.

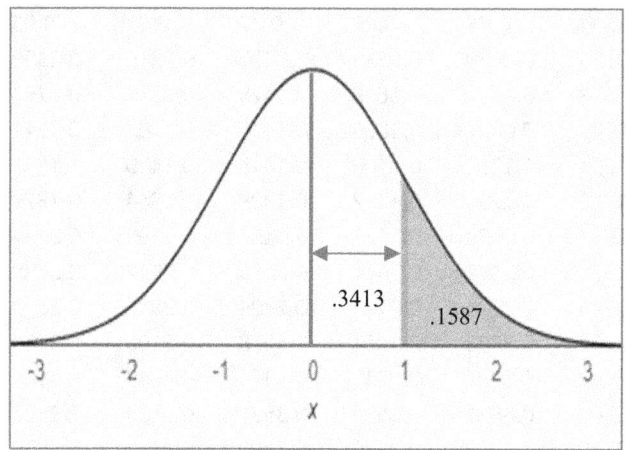

Knowing that the each half of the curve is .5, we can now use the value we found in the table to compute the area of the curve represented by the shading.
$$= .5 - .3413 = .1587$$

So, the probability of finding a data point with a value higher than 110 is 15.87%.

So, let's approach this another way.

What is the probability of finding a value data point with a value lower than 110?

We still use the area we located in the table, .3413, but that is still not the complete answer. Because we are looking for any value lower than 110, the value could also be lower than the mean. To account for this, we must include the area from the mean to the z score of 1, and the entire half of the curve to the left of the mean.

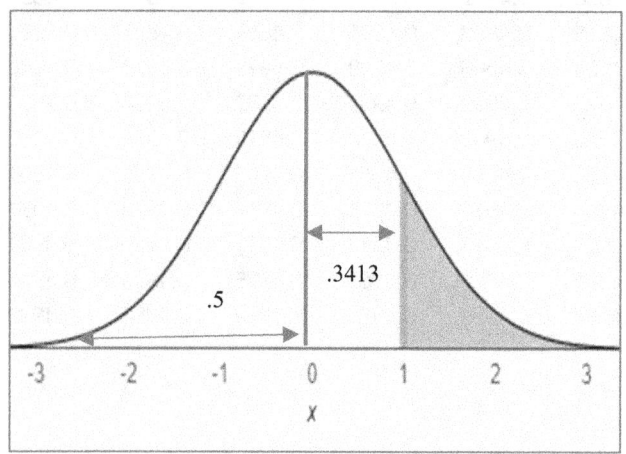

So, the probability of finding a data point with a score lower than 110 is:

$$.5 + .3413 = .8413$$

Now, what if we wanted to know the probability of finding a data point with a value between 90 and 110?

The z scores are: $(110 - 100) / 10$ and $(90 - 100) / 10$
$= 1$ and -1

While some normal curve probability tables will display areas for negative z scores, some, like the example on the previous page, do not. If the table you are using does not have values specifically for negative z scores, you can use the absolute value of the z score 1, for -1, and use the value associated for that value, but continuing to realize the score represent a position in the curve to the left of (less than) the mean.

Surviving Statistics

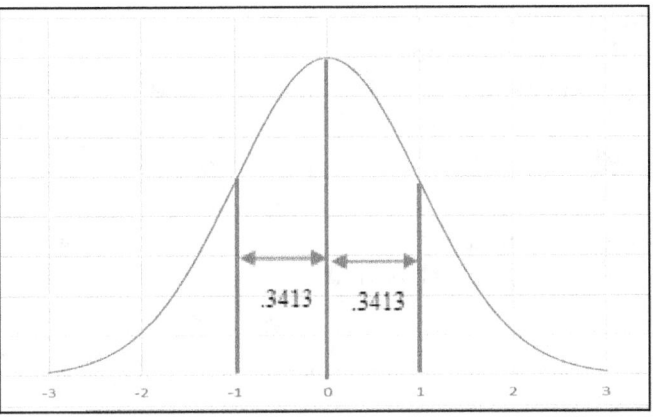

The probability of finding a data point with a value between 90 and 110 is:
.3413 + .3413 = .6826

Solving for *x*

Sometimes you will be given an area or probability and then be asked to find the value that relates to that probability. For example, Oliver's Outboards makes and sells outboard motors for boats. After several years in business, Oliver knows that his outboards fail, on average, after 500 hours of use with a standard deviation of 22 hours. (Oliver is a data geek.) Oliver wants to offer his customers a warranty but is willing to have only 2% of his motors fail during the warranty period. What warranty, in running hours, should Oliver offer on his outboards?

We are solving for x in this case, the number of hours. The steps in doing this are:
1. Find the z score associated with the top 2%.

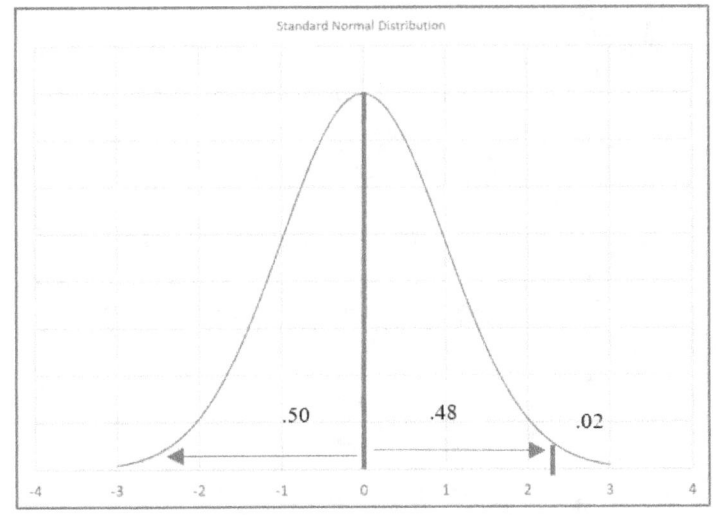

Using the table, we locate a z score for the area of .48.

The z score is between 2.05 and 2.06. Using the table, we can approximate with 2.055. Using Excel or the formula, we compute an actual value of 2.054. We will use 2.055 in our computations.

2. After finding, z, use algebra with the z score formula to solve for x.
   That formula becomes:
   $$x = \mu + or - z(\sigma)$$

Surviving Statistics

Add if the value is higher than the mean and subtract if less. Since Oliver wants 98% to last longer than the warranty period, this value will be less than the mean.

$$x = 500 - 2.054(22) = 454.812$$

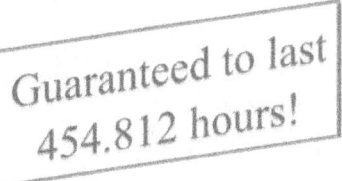

So, Oliver should offer a warranty close to 454.812 hours to ensure that only 2% or less of his motors fail during the warranty period.

Try it in Excel:

Excel's Norm.Dist() function will compute the area under the curve for a given value. For this to work properly, you will turn on the Cumulative argument. However, this causes Excel to add .5 to the area because it includes both halves of the curve. The images show Excel's solution.

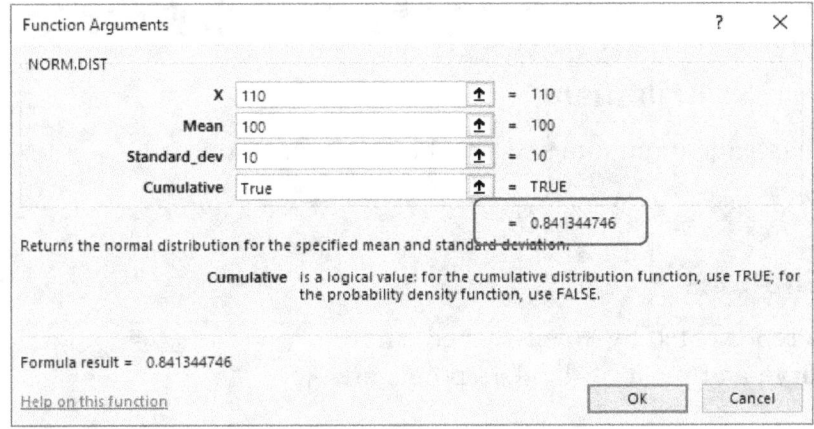

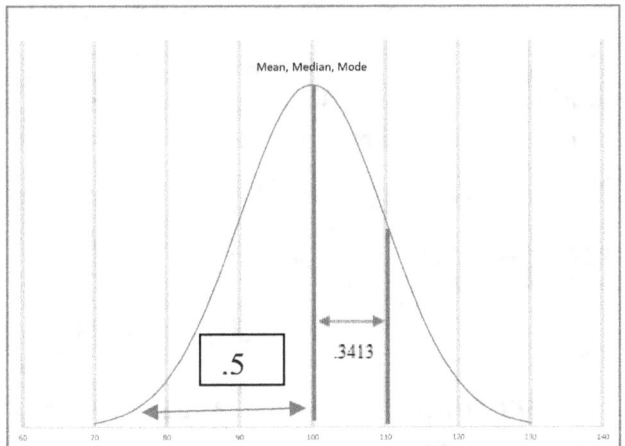

.84134476 - .5 = .34134

In situations where you need to solve for the x value, you can use Excel's NormInv() function.

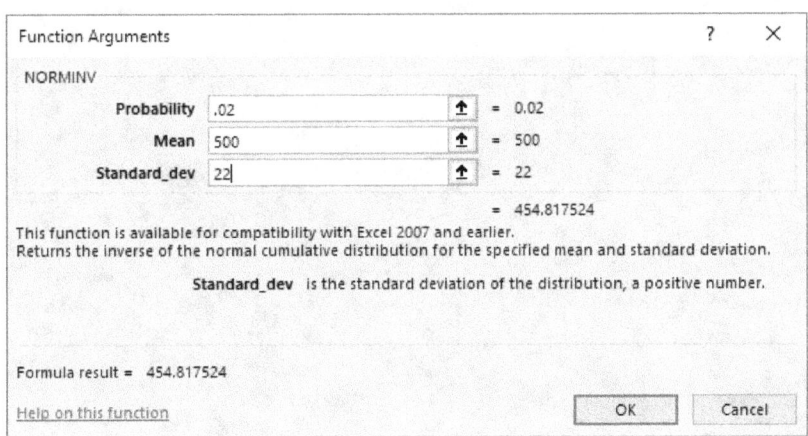

Chapter Highlights

- **Uniform Distributions**

With a uniform distribution, the probably of any outcome is equal to that of any other. To illustrate this, assume you have arrived at a bus stop. Uniform probability distributions are represented as a rectangle with the minimum and maximum values.

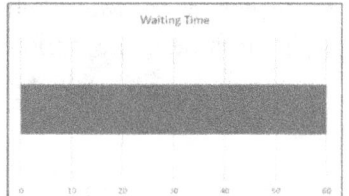

  - **The mean of a uniform distribution:**

$$\mu = \frac{a+b}{2}$$ where a is the minimum and b is the maximum value

- **Normal Probability Distributions**

A normal probability distribution is represented by a bell-shaped curve. The mean is highest point in the curve and most of the data points are distributed closely around the mean.

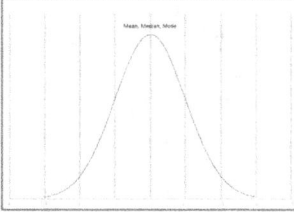

- **Standard Normal Distribution (Z scores)**

A z score equates to the number of standard deviations a value is from the mean. Knowing the z score for a value, we can then compute the area under the curve, or probability.

  - **The formula for computing a z score is:**

$$z = \frac{x - \mu}{\sigma}$$

- **To find the probability associated with a z score, use the normal distribution table.**

Chapter 8: Sampling and Sampling Distributions

This chapter will cover:
- **Sampling Methods**
- **Sampling Error**
- **Sampling Distribution of the Sample Mean**
- **Central Limit Theorem**
- **Computing a z score from a Sample Mean**

Chapter 8: Sampling and Sampling Distributions

We have already discussed the differences between a sample and a population and statistics and parameters. We have discussed several reasons why we use samples instead of populations. In this chapter we will discuss how to properly select a sample. We will also show why we can trust samples to be representative of a population.

Sampling methods

Some of the sampling methods you should be aware of include:

SIMPLE RANDOM SAMPLING

To ensure the sample represents the population, every object or person in the population should have an equal chance of being selected. With simple random sampling, selection of any one object to be included in the sample occurs at random. No one item or person, or group of persons or items, is selected beforehand or targeted to be sampled. Practical examples of this include drawing a name out of a hat. If all names have an equal chance of being selected, then the name drawn was completely random.

If you decide to use this method, you can use a table of random numbers which are usually included in textbooks or available online. You can also use Excel to generate random numbers as well.

SYSTEMATIC RANDOM SAMPLING

With systematic random sampling, you take a list of the population you want to sample and then take every k^{th} item. For example, in a list of 100 employees, you may decide to survey every 5th employee.

Sampling error

In most cases, properly collected and sized samples are representative of the population. We will understand why we can trust this statement when we discuss sampling distributions shortly. However, because the sample is not a perfect representation of the population, there will be a difference between the statistics computed from the sample and the parameters computed from the entire population.

We refer to the difference between the sample statistic and the population parameter as the sampling error.

Sampling error – difference between statistic and parameter	
Statistic	Parameter
$\bar{x}$	μ
s^2	σ^2
s	σ

Sampling Distribution of the Sample Mean

You should recall that a probability distribution is a listing of all possible outcomes and the probability of each outcome. Similarly, a sampling distribution of the sample mean is a listing of all possible sample means you can generate from a particular sample size, n.

To illustrate a sampling distribution of the sample mean, assume a certain statistics class has 10 students. You want to know how your own GPA compares with the average GPA of the class as a whole. So, you have decided to select two students at random and compute the mean GPA of those two students. Each student in the class and their GPA is shown below.

Student	GPA
Ahmad	3.80
Alicia	2.30
Bob	2.50
Charles	3.95
Kathy	3.00
Maria	3.75
Min	3.30
Thomas	2.40
William	3.10
Xavier	3.00
Mean	3.11

Is the mean of the two-student sample an accurate representation of the population (class) mean? Let's see.

With a sample size, n, of 2 and a population, N, of 10, there are 45 possible samples of two student each. We use the combination formula to tell us home many possible combinations were possible.

$$_nC_r = \frac{n!}{r!(n-r)!}$$

The sampling distribution of the sample mean lists all possible samples of two and the computed mean of each possible sample.

The following table shows every possible two-student combination, 45 in all, and the mean of that sample. The difference between the sample mean and the population mean is the sampling error. Notice that the mean of the sample means is exactly equal to the population mean.

Also note the sum the sampling error in the distribution is zero. **The important concept here is that samples can accurately represent the population.**

						Sample mean	Sampling error
1	Ahmad	3.80	Alicia	2.30	3.05	-0.06
2	Ahmad	3.80	Bob	2.50	3.15	0.04
3	Ahmad	3.80	Charles	3.95	3.88	0.77
4	Ahmad	3.80	Kathy	3.00	3.40	0.29
5	Ahmad	3.80	Maria	3.75	3.78	0.67
6	Ahmad	3.80	Min	3.30	3.55	0.44
7	Ahmad	3.80	Thomas	2.40	3.10	-0.01
8	Ahmad	3.80	William	3.10	3.45	0.34

Surviving Statistics

9	Ahmad	3.80	Xavier	3.00	3.40	0.29	
10	Alicia	2.30	Bob	2.50	2.40	-0.71	
11	Alicia	2.30	Charles	3.95	3.13	0.01	
12	Alicia	2.30	Kathy	3.00	2.65	-0.46	
13	Alicia	2.30	Maria	3.75	3.03	-0.09	
14	Alicia	2.30	Min	3.30	2.80	-0.31	
15	Alicia	2.30	Thomas	2.40	2.35	-0.76	
16	Alicia	2.30	William	3.10	2.70	-0.41	
17	Alicia	2.30	Xavier	3.00	2.65	-0.46	
18	Bob	2.50	Charles	3.95	3.23	0.12	
19	Bob	2.50	Kathy	3.00	2.75	-0.36	
20	Bob	2.50	Maria	3.75	3.13	0.01	
21	Bob	2.50	Min	3.30	2.90	-0.21	
22	Bob	2.50	Thomas	2.40	2.45	-0.66	
23	Bob	2.50	William	3.10	2.80	-0.31	
24	Bob	2.50	Xavier	3.00	2.75	-0.36	
25	Charles	3.95	Kathy	3.00	3.48	0.37	
26	Charles	3.95	Maria	3.75	3.85	0.74	
27	Charles	3.95	Min	3.30	3.63	0.52	
28	Charles	3.95	Thomas	2.40	3.18	0.06	
29	Charles	3.95	William	3.10	3.53	0.42	
30	Charles	3.95	Xavier	3.00	3.48	0.37	
31	Kathy	3.00	Maria	3.75	3.38	0.27	
32	Kathy	3.00	Min	3.30	3.15	0.04	
33	Kathy	3.00	Thomas	2.40	2.70	-0.41	
34	Kathy	3.00	William	3.10	3.05	-0.06	
35	Kathy	3.00	Xavier	3.00	3.00	-0.11	
36	Maria	3.75	Min	3.30	3.53	0.42	
37	Maria	3.75	Thomas	2.40	3.08	-0.04	
38	Maria	3.75	William	3.10	3.43	0.32	
39	Maria	3.75	Xavier	3.00	3.38	0.27	
40	Min	3.30	Thomas	2.40	2.85	-0.26	
41	Min	3.30	William	3.10	3.20	0.09	
42	Min	3.30	Xavier	3.00	3.15	0.04	
43	Thomas	2.40	William	3.10	2.75	-0.36	
44	Thomas	2.40	Xavier	3.00	2.70	-0.41	
45	William	3.10	Xavier	3.00	3.05	-0.06	
				Mean	3.11		
					Sum	0.00	

The Central Limit Theorem

The central limit theorem states that a sampling distribution of the sample means can be approximated by a normal distribution. The larger the sample size, the closer to a normal distribution the sampling distribution of the sample mean becomes.

Even with a sample size of two, as in the class GPA example, you can see the central limit theorem expressed. The first illustration is the frequency distribution to the population GPA scores. For ease of interpretation, the scores are grouped with a width of .2.

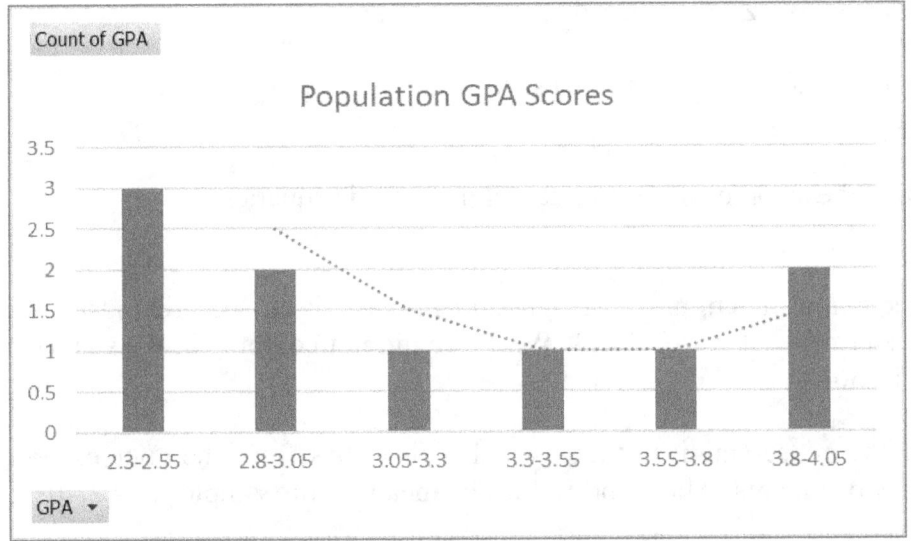

The second illustration is a frequency distribution of the sampling distribution of the sample means for a sample size of two. You should notice that even with an extremely small sample size, the distribution begins to approximate a normal distribution.

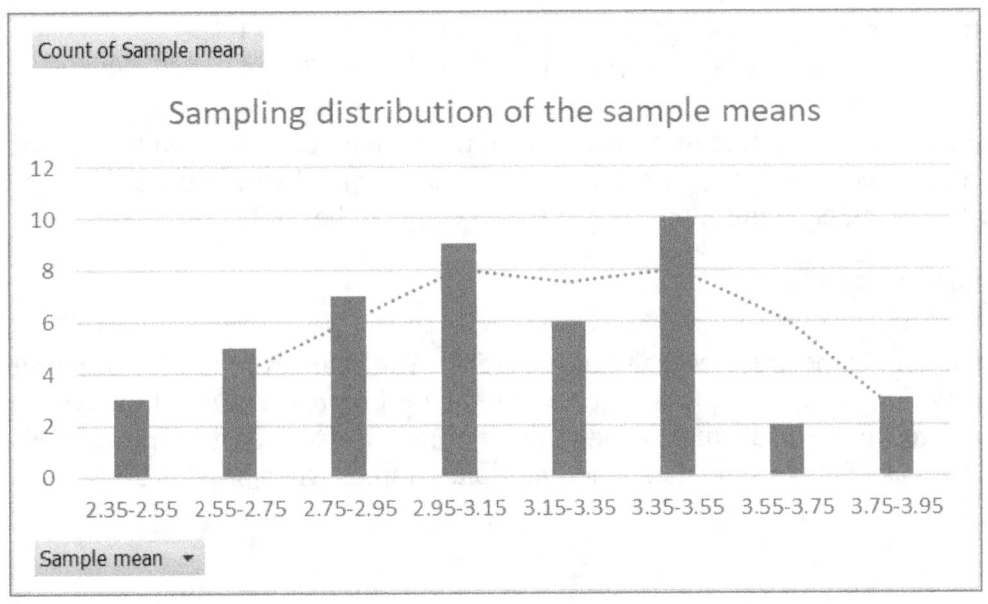

So, what is the big deal about the central limit theorem? It lets us assume that our sample data can be represented with a normal distribution. This allows us to use sample means and compute z scores. This, in turn, lets us compute probabilities or areas under the curve with the z scores computed from our sample means. As you progress through statistics, you will often see the words "assume the data is normally distributed" in practice problems and examples. We can make this assumption based on the central limit theorem.

Using a Sample Mean to Compute a Z score

Up to this point we have computed z scores using one value, x. Since statistics is concerned with samples, we will also want to compute z scores based on sample means.

Here is the formula we use to compute a z score from a sample mean.

$$z = \frac{\bar{x} - \mu}{\sigma/\sqrt{n}}$$ Notice that the sample size is an integral part of this equation.

Here is an example of this equation in action. In the ten-student class discussed previously, we computed the population mean of the GPA to be 3.11. While we did not compute it, the standard deviation of this population is .5643.

Assume we wanted to find two representative students and their GPAs from the course. We randomly select a sample of two students, Alicia and Bob. The mean of this sample, $\bar{x}$, is 2.40.

Are these two students a good representation of the class as a whole?

Or, since the sample mean is less than the population mean, what is the probability of finding a sample with this GPA or lower?

Using the z score formula we solve for z as: (2.40 – 3.11) / (.5643/1.414) = -1.78
1.414 is the square root of the sample size, 2.

The z score is negative because the sample mean is less than the population mean. Now, we find the probability (area), associated with a z score of 1.78. Using the normal distribution table which we have placed on the following page in Figure 28, we find the area to be .4625.

How do we interpret this?

The chances of finding a sample mean lower than this is 3.75%. The chances of finding a sample mean higher than 2.4 is 96.25%. The sample we took is not the best representation of the typical students in the course. Here we see an example of sampling error. We also see why larger sample sizes are important, as well as why we should conduct multiple samples when possible.

Even though the two students we selected at random do not perfectly represent the typical students, if we sampled enough students, we would arrive at a mean GPA very close to the population mean GPA.

z	0.00	0.01	0.02	0.03	0.04	0.05	0.06	0.07	0.08	0.09
0	0.0000	0.0040	0.0080	0.0120	0.0160	0.0199	0.0239	0.0279	0.0319	0.0359
0.1	0.0398	0.0438	0.0478	0.0517	0.0557	0.0596	0.0636	0.0675	0.0714	0.0753
0.2	0.0793	0.0832	0.0871	0.0910	0.0948	0.0987	0.1026	0.1064	0.1103	0.1141
0.3	0.1179	0.1217	0.1255	0.1293	0.1331	0.1368	0.1406	0.1443	0.1480	0.1517
0.4	0.1554	0.1591	0.1628	0.1664	0.1700	0.1736	0.1772	0.1808	0.1844	0.1879
0.5	0.1915	0.1950	0.1985	0.2019	0.2054	0.2088	0.2123	0.2157	0.2190	0.2224
0.6	0.2257	0.2291	0.2324	0.2357	0.2389	0.2422	0.2454	0.2486	0.2517	0.2549
0.7	0.2580	0.2611	0.2642	0.2673	0.2704	0.2734	0.2764	0.2794	0.2823	0.2852
0.8	0.2881	0.2910	0.2939	0.2967	0.2995	0.3023	0.3051	0.3078	0.3106	0.3133
0.9	0.3159	0.3186	0.3212	0.3238	0.3264	0.3289	0.3315	0.3340	0.3365	0.3389
1	0.3413	0.3438	0.3461	0.3485	0.3508	0.3531	0.3554	0.3577	0.3599	0.3621
1.1	0.3643	0.3665	0.3686	0.3708	0.3729	0.3749	0.3770	0.3790	0.3810	0.3830
1.2	0.3849	0.3869	0.3888	0.3907	0.3925	0.3944	0.3962	0.3980	0.3997	0.4015
1.3	0.4032	0.4049	0.4066	0.4082	0.4099	0.4115	0.4131	0.4147	0.4162	0.4177
1.4	0.4192	0.4207	0.4222	0.4236	0.4251	0.4265	0.4279	0.4292	0.4306	0.4319
1.5	0.4332	0.4345	0.4357	0.4370	0.4382	0.4394	0.4406	0.4418	0.4429	0.4441
1.6	0.4452	0.4463	0.4474	0.4484	0.4495	0.4505	0.4515	0.4525	0.4535	0.4545
1.7	0.4554	0.4564	0.4573	0.4582	0.4591	0.4599	0.4608	0.4616	0.4625	0.4633
1.8	0.4641	0.4649	0.4656	0.4664	0.4671	0.4678	0.4686	0.4693	0.4699	0.4706
1.9	0.4713	0.4719	0.4726	0.4732	0.4738	0.4744	0.4750	0.4756	0.4761	0.4767
2	0.4772	0.4778	0.4783	0.4788	0.4793	0.4798	0.4803	0.4808	0.4812	0.4817
2.1	0.4821	0.4826	0.4830	0.4834	0.4838	0.4842	0.4846	0.4850	0.4854	0.4857
2.2	0.4861	0.4864	0.4868	0.4871	0.4875	0.4878	0.4881	0.4884	0.4887	0.4890
2.3	0.4893	0.4896	0.4898	0.4901	0.4904	0.4906	0.4909	0.4911	0.4913	0.4916
2.4	0.4918	0.4920	0.4922	0.4925	0.4927	0.4929	0.4931	0.4932	0.4934	0.4936
2.5	0.4938	0.4940	0.4941	0.4943	0.4945	0.4946	0.4948	0.4949	0.4951	0.4952
2.6	0.4953	0.4955	0.4956	0.4957	0.4959	0.4960	0.4961	0.4962	0.4963	0.4964
2.7	0.4965	0.4966	0.4967	0.4968	0.4969	0.4970	0.4971	0.4972	0.4973	0.4974
2.8	0.4974	0.4975	0.4976	0.4977	0.4977	0.4978	0.4979	0.4979	0.4980	0.4981
2.9	0.4981	0.4982	0.4982	0.4983	0.4984	0.4984	0.4985	0.4985	0.4986	0.4986
3	0.4987	0.4987	0.4987	0.4988	0.4988	0.4989	0.4989	0.4989	0.4990	0.4990

Area under the normal curve

Figure 28: Normal distribution table, z score 1.78

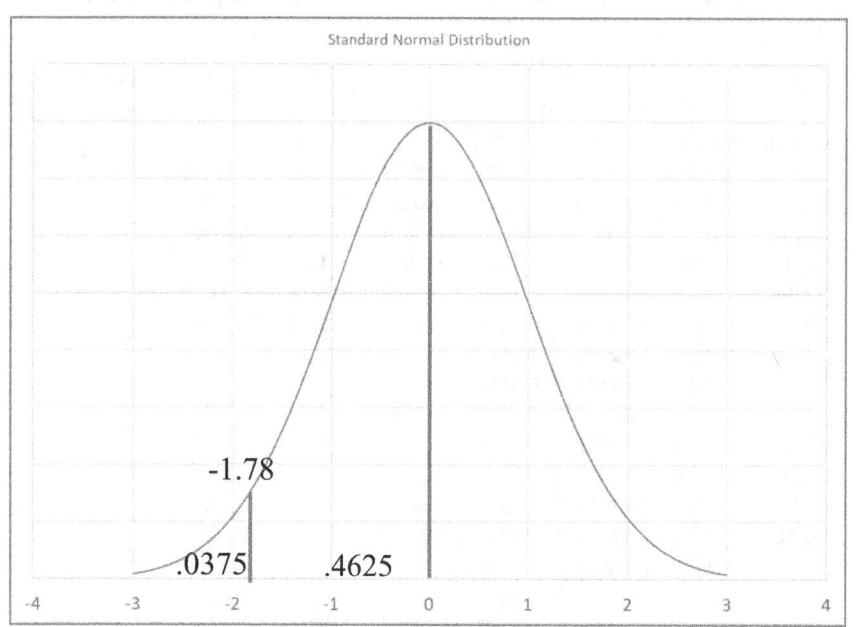

Standard Error of the Mean

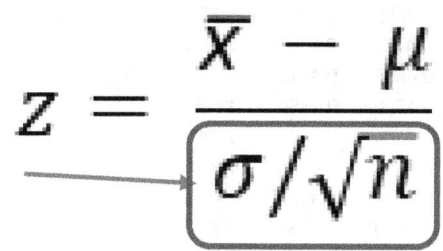

The denominator of the formula you just used to compute a z score based on a sample mean is called the standard error of the mean, or more accurately, the standard deviation of the sampling distribution of the sample mean.

You will use the standard error of the mean as a stand-alone computation. You will also use other upcoming computations, such as confidence intervals, which we will discuss in the next chapter.

Chapter Highlights

- **Sampling Methods**

Some of the sampling methods you should be aware of include:

  - **Simple Random Sampling**

    Selection of any one object occurs at random. No one item or person, or group of persons or items, is selected beforehand and all have an equal chance of being included in the sample.

  - **Systematic Random Sampling**

    Systematic random sampling takes k^{th} item from a list.

- **Sampling Error**

The difference between the sample statistic and the population parameter is called the sampling error.

Sampling error – difference between statistic and parameter	
Statistic	Parameter
$\bar{x}$	μ
s^2	σ^2
s	σ

- **Sampling Distribution of the Sample Mean**

A sampling distribution of the sample mean is a listing of all possible sample means that can be generated from a sample size of *n*.

- **The Central Limit Theorem**

The central limit theorem states that a sampling distribution of the sample mean can be approximated by a normal distribution. The larger the sample size these, the closer to a normal distribution the sampling distribution becomes.

- **Computing a z Score from a Sample Mean:** $\quad z = \dfrac{\bar{x} - \mu}{\sigma/\sqrt{n}}$

- **Standard error of the mean:** $\quad z = \dfrac{\bar{x} - \mu}{\sigma/\sqrt{n}}$

Surviving Statistics

Chapter 9: Confidence Intervals and Point Estimates

This chapter will cover:

- **Point Estimate**
- **Confidence Interval**
- **Margin of Error**
  - **Known Population Standard Deviation**
  - **Unknown Population Standard Deviation**
  - **Proportions**
- **Choosing a Sample Size**

Chapter 9: Confidence Intervals and Point Estimates

Point Estimate

Based on what we learned about the sampling distribution of the sample means, if we want to estimate parameter values for a population, the best estimate of a population value is a statistic computed from a sample. If we wanted to estimate the mean GPA of the population of ten students, we could take a sample and compute the mean. That statistic, the sample mean, would be the best estimate of the population mean, assuming we did not have the values for the entire population.

However, while values derived from samples are the best estimates, we also know that a mean computed from a sample may not be exact due to sampling error. As we saw in the last chapter with the sampling distribution of the sample means, some sample means had a larger sampling error than other samples.

Confidence Interval

Because a statistic computed from a sample may not be exact, we use confidence intervals. A confidence interval is a range of values that we are confident contains the population parameter. We refer to our level of certainty, or confidence, as the confidence interval.

Confidence intervals are typically between 90% and 99%. So, using a 95% confidence level, you can be 95% confident that the true population parameter will fall within the confidence interval, the range of values less than and greater than the value we derived from the sample.

MARGIN OF ERROR – KNOWN POPULATION STANDARD DEVIATION

Confidence intervals use a margin of error. After computing the margin of error, we add and then subtract it from the sample mean to create the confidence interval (range). When we know the population standard deviation, we compute a margin of error based on a z score.

To understand this concept, let's assume we want to accurately predict the population GPA for the ten-student class using a two-student sample. The sample we randomly selected was Ahmad and Kathy and the sample mean is 3.0

We want to be 95% confident in estimating the true population mean, so we choose a 95% confidence interval. We will create a range of values, centered around the sample mean of 3.0. Once we compute that range of values, we will have 95% confidence that the population mean lies within that range.

We know the population standard deviation is .5643. The formula for computing a margin of error from a sample mean is:

$$\text{Margin of error} = z \frac{\sigma}{\sqrt{n}}$$

Notice that the margin of error formula is determined by the z score corresponding to the confidence level multiplied by the standard error of the mean.

The confidence interval equals:

$$\bar{x} \pm z \frac{\sigma}{\sqrt{n}}$$

We have the sample mean, 3.0, the sample size, 2, and the population standard deviation, .5643. Now, we must find the z score associated with a 95% confidence level. This z score will include 95% of the area under the normal curve.

Because the confidence interval will be a range that is smaller and larger than the mean, we are looking for a two-tail z score. For a 95% confidence level, we are looking for a z score that represents 5% of the area under the curve. However, because we are adding and subtracting the margin of error to the sample mean, we will divide the 5% of the area not under the curve by two. Therefore, we are looking for a z score that represents 47.5% of the area between the population mean and the z score.

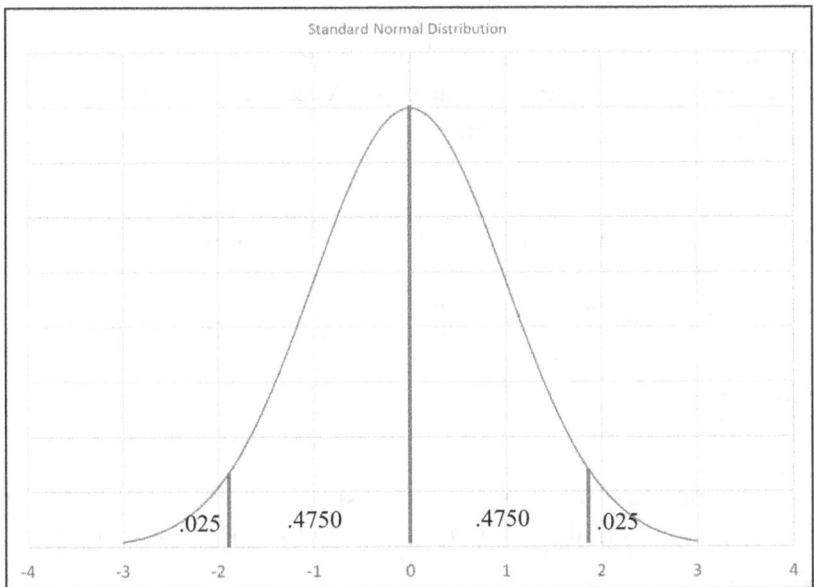

Figure 29: A 95% confidence interval uses both tails

Using the normal distribution table, z scores, we locate the area of .4750 in the normal distribution table such as the one in Figure 28 on page 93, and find that area equates to the z score of 1.96.

Now we can compute the confidence interval based on the sample mean.

Margin of error = 1.96(.564/1.41) = .782

We add and subtract the margin of error to our sample mean and get a 95% confidence interval of 2.218 to 3.782.

We can be 95% confident that the true population mean (parameter) is between 2.218 and 3.782.

MARGIN OF ERROR – UNKNOWN POPULATION STANDARD DEVIATION

The previous formula for computing the margin of error assumes we know the standard deviation of the population, σ. If you are conducting new research on samples, it is unlikely you will have parameters such as the mean and standard deviation of the population. Although this may seem to create a problem because z scores and confidence intervals all require the population standard deviation. Statisticians have solved the problem by using another distribution.

The *t* distribution is similar to the *z*'s normal curve but is flatter and contains more information in the tails. The data in a *t* distribution is more dispersed from the mean than in a normal (z) distribution. We can compute *t* scores using the sample standard deviation, *s*, rather than the population standard deviation. Unlike z scores, *t* scores vary depending on the sample size. The *t* distribution accounts for the sample size in *degrees of freedom.*

The degrees of freedom for computing confidence intervals using a *t* score is **n – 1**.

Let's see how this works with an example. Assume the GPAs from the ten-student course is now a sample of all statistic students in the college (the population). We want to create a 95% confidence interval for the means of the students in the college.

Student	GPA
Ahmad	3.80
Alicia	2.30
Bob	2.50
Charles	3.95
Kathy	3.00
Maria	3.75
Min	3.30
Thomas	2.40
William	3.10
Xavier	3.00
Mean	**3.11**

We will now consider the mean of these students as a sample mean, $\bar{x}$.

The sample standard deviation, *s,* is .5948.

$$Margin\ of\ error = t\frac{s}{\sqrt{n}}$$

The formula to compute the margin of error is similar to the one used for z, except it uses the sample standard deviation and a critical *t* value, not *z*.

Since we have the standard deviation and the sample size, the next step is to look up the *t* score in the *t* distribution table.

As previously mentioned, the *t* distribution uses degrees for freedom, which in this case will be *n – 1*, or 9 for this sample of 10 students.

A confidence interval includes values both greater than and less than the mean, so the *t* score will also be a two-tailed, just like the z score was.

Some *t* distribution tables display columns for both one and two tail tests. This may actually be more confusing than helpful. Either way, the column we want is the .05 two tail or .025 one tail (2.5% in each tail). After locating the degrees of freedom row and the confidence level column, you can then locate the *t* score, which in this case is 2.262 as you can see in Figure 30.

	Partial *t* distribution table					
	One tail					
	0.1	0.05	0.025	0.01	0.005	0.0005
	Two tail					
DF	0.2	0.1	0.05	0.02	0.01	0.001
1	3.078	6.314	12.706	31.821	63.657	636.619
2	1.886	2.920	4.303	6.965	9.925	31.599
3	1.638	2.353	3.182	4.541	5.841	12.924
4	1.533	2.132	2.776	3.747	4.604	8.610
5	1.476	2.015	2.571	3.365	4.032	6.869
6	1.440	1.943	2.447	3.143	3.707	5.959
7	1.415	1.895	2.365	2.998	3.499	5.408
8	1.397	1.860	2.306	2.896	3.355	5.041
9	1.383	1.833	2.262	2.821	3.250	4.781
10	1.372	1.812	2.228	2.764	3.169	4.587
11	1.363	1.796	2.201	2.718	3.106	4.437
12	1.356	1.782	2.179	2.681	3.055	4.318
13	1.350	1.771	2.160	2.650	3.012	4.221
14	1.345	1.761	2.145	2.624	2.977	4.140
15	1.341	1.753	2.131	2.602	2.947	4.073

Figure 30: A sample *t* distribution table

You should notice that the *t* score for the 95% confidence, 2.262 is larger than the *z* score for the same level, which was 1.96. *T* scores are larger because the curve is flatter than a *z* distribution, which, again, means there is more dispersion from the mean.

The margin of error for a 95% confidence interval using the *t* distribution is:

$$2.262((.5643/1.414) = .903$$

The 95% confidence interval is: (2.207, 4.013). We can be 95% confident that the population mean falls within this range.

Z OR T?

If you have the population standard deviation, use the *z* distribution. If not, use the *t* distribution.

CONFIDENCE INTERVALS FOR PROPORTIONS

Proportions are fractions, ratios, or percentages of the whole that possess a specific trait, attribute, or interest. Examples could include the percentage of statistics students who failed the course compared to all statistic students we sampled. Or the proportion of products coming off the assembly line which failed final inspection compared to all products we sampled that day.

Proportions rely on the nominal measurement scale and there are only two possible outcomes. Often these are attributes that can be answered with a yes or no, or pass or fail. The formula to compute the sample proportion is:

$$p = \frac{x}{n}$$

x is the number with the attribute (answered yes, passed, etc....)
n is the sample size

For our example, we will assume we sampled 100 moviegoers and asked each if they liked the last movie they watched. Forty of our sampled moviegoers answered "yes". The sample proportion is 40/100 or .4.

We want to compute the 95% confidence interval for this sample proportion.

The formula to compute a confidence interval for a sample proportion is:

$$p \pm z \sqrt{\frac{p(1-p)}{n}}$$

Notice that proportions use the z statistic, even though we do not have the population standard deviation. We know from past examples, that the z score for a 95% confidence level, two-tail, is 1.96. So, we compute the margin of error as:

$$p \pm 1.96 \sqrt{\frac{.4(1-.4)}{100}}$$

The margin of error is .096, so the confidence interval becomes: **.304 to .496.**

This means we can state, with 95% confidence that the proportion of movie goers who liked the last move they saw falls between .304 and .496.

Surviving Statistics

CHOOSING A SAMPLE SIZE

As we have seen with many computations already, the sample size is a very important factor. We also saw with the central limit theorem that the larger the sample size, the more the sampling distribution of the sample means approaches a normal distribution curve.

So what is the optimal sample size?

Computing the correct sample size depends on three variables:
1. The maximum acceptable margin of error.
2. The desired level of confidence.
3. The dispersion of the variable being studied in the population (standard deviation).

Sample Size for Population Mean

When you are attempting to determine the proper sample size, you will use one formula when you are measuring means and another for preparations. We will begin with the formula for choosing a sample size for estimating population mean values.

The formula for computing a sample size is:

$$n = \left(\frac{z\sigma}{E}\right)^2$$

In this formula, E is allowable margin of error. We'll discuss this in the example.

As an example, assume you want to estimate the GPA of all the students at a particular college taking statistics. You do not want to survey the entire population, so you decide to survey a sample. Assuming you want a 95% confidence level, and are only willing to accept a margin of error of 0.1, or 10%, how many students do you need to survey?

You should have noticed that we are missing one required variable to compute the sample size, the population standard deviation. While we could probably get this information from the college's administration, we will assume they have very bad records, or are unwilling to assist in our research, so we cannot procure that value.

As we have already discussed, when you are doing your own research, you will often be unable to procure values relating to the population, such as the standard deviation. So, does that mean we cannot compute a sample size? Fortunately, there are ways we can estimate this unknown value.

Ways to estimate the population standard deviation include:
1. Use data from previous studies, or conduct a pilot study
2. Estimate the standard deviation from the population range.
   If you know the possible range, you can then divide the range by 4 or 6, to estimate the population standard deviation.

Why either 4 or 6?

If the population range is correct, it will include all possible values. Recall that the empirical rule tells us that 99.7% of the values fall within +/- three standard deviations of the mean, which totals six. The empirical rule also tells us that 94.7% of the values fall within +/- two standard deviations of the mean, which totals four.

For our example, we will divide the rage by six.

For our example, GPA's can range from 0 to 4. The range is four. Our estimated population standard deviation is 4/6 = .667

Now we can compute the correct sample size: $n = \left(\dfrac{1.96(.667)}{.1}\right)^2 = 171$

The *z* score of 1.96 represents the 95% confidence interval, 2.5% in each tail.

We have computed a required sample size of 171. This is much larger than the ten-student sample size we have been using. A sample size of 10 would be the appropriate size if we allowed a margin of error of .4.

Sample Size for Proportions

As we have already discussed, we do not use standard deviation when we compute statistics for proportions. The sample size for proportions uses the population proportion and the margin of error.

$n = p^*(1 - p^*)\left(\dfrac{Z}{E}\right)^2$ In this formula, p^* is the population proportion. If you do not know this, you can estimate it. We can estimate the population proportion by:

1. Conducting a pilot study
2. Using previous studies
3. Using your "best guess"
4. Using 0.5 as an estimate

We will continue to work in proportions in upcoming chapters.

Chapter Highlights

- **Point Estimate**

The best estimate of a population parameter is a sample statistic.

- **Confidence Interval**

The confidence we have, 90% to 99%, that a population parameter will fall within a given range of values generated from a sample or samples. A confidence interval will be the statistic, such as mean, +/- the margin of error.

  - **Margin of Error – Known Population Standard Deviation** $\quad z\dfrac{\sigma}{\sqrt{n}}$

  - **Margin of Error – Unknown Population Standard Deviation** $\quad t\dfrac{s}{\sqrt{n}}$

  - **Confidence Intervals for Proportions** $\quad p \pm z\sqrt{\dfrac{p(1-p)}{n}}$

- **Choosing a Sample Size**

  - **Sample Size for Population Mean** $\quad n = \left(\dfrac{z\sigma}{E}\right)^2$

  - **Sample Size for Proportions** $\quad n = p^*(1-p^*)\left(\dfrac{z}{E}\right)^2$

Chapter 10: Hypothesis Testing: One Sample

This chapter will cover:

    Five-step Hypothesis Testing Procedure

        One Sample Known Population Standard Deviation

        One Sample Unknown Population Standard Deviation

        One Proportion

Chapter 10: Hypothesis Testing: One Sample

You might recall from our first chapter that inferential statistics attempts to estimate, generalize, or predict something about a population. We introduced the concept of inferential statistics in the first few chapters. Then, we dug deeper into inferential statistics with the discussion of z scores and areas under the curve. Hypothesis testing is the formal process of checking the assumptions or inferences we have made. Researchers use hypothesis testing to test existing theories, propose new ones, test the effectiveness of a treatment or intervention, or just to compare two groups of observations. Hypothesis testing can also be used in quality control, such as checking the accuracy of weights and contents.

Hypothesis testing is a foundation of statistical research. From this point on, hypothesis testing will become an integral part of almost everything we do in this text with statistics.

Five Step Hypothesis Testing Procedure

Hypothesis testing can be a difficult concept to grasp. However, it really is not very difficult, especially if you follow the five-step hypothesis testing procedure in this chapter. This chapter will focus on hypothesis testing when we are comparing only one sample with a population.

Once you master the five step process we introduce in this chapter, you will use it with many additional statistical computations. Several additional chapters of this text will rely on this same five-step procedure.

1. ### STATE THE NULL AND ALTERNATE HYPOTHESIS

Hypothesis testing requires two hypotheses, the null and the alternate. At the end of the hypothesis testing procedure, you will reject or not reject the null hypothesis. If you reject the null, then you may be able to accept the alternate. Here is an explanation of the null and alternate hypotheses:

The Null Hypothesis
The null hypothesis, H_0, is what is assumed to be true, the status quo, "common knowledge," or an established theory. In statistics, the null hypothesis is always expressed with some form of equality, such as <=, >=, or =.

For example, if a bag of potato chips states that it contains 8.7 ounces of chips, then the null hypothesis would be: chips in >= 8.7 ounces.

We include greater than 8.7 because to ensure there are no lawsuits, the production company will attempt to error on the side of too many chips, not too few.

The Alternate Hypothesis
The alternate hypothesis, H_1 is mathematically contrary to the null hypothesis. Since the null hypothesis, H_0, contains an equality, the alternate hypothesis will contain not equal to (<>), greater than (>), or less than (<).
Example #1:

Assume you oversee production at a potato chip company. Each bag is labeled that is has 8.7 ounces of chips. You need to constantly monitor the chip dispenser to ensure the bags contain at least 8.7 ounces of chips. Underfilling the bags would result in customer complaints and even, potentially, lawsuits. Because you think there may be a problem with the chip dispenser you develop the following hypotheses:

Null, H_0: Chip weight >= 8.7 ounces.

Alternate, H_1: Chip weight < 8.7 ounces.

Note: *You are only concerned about underfilling the bags with this example. In the "real world" over filling the bags would also be a problem. While overfilling the bags would make your customers happy, you may not stay in business very long by giving away your product.*

Without making this too complex, the solution businesses use is related to the confidence interval you learned about in chapter 9. You may, for example allow a margin of error of .1 ounces. Your chip dispenser might shoot for 8.8 ounces and is working properly (within tolerance) if it dispenses from 8.7 to 8.9 ounces.

Example #2:
You are testing a new procedure for making sandwiches at a fast food restaurant and believe your new method is faster and set out to test your theory. The null and alternate hypotheses would be stated as:

H_0: Mean assembly time with new procedure >= mean assembly time with old procedure

H_1: Mean assembly time with new procedure < mean assembly time with old procedure

Example #3
You are testing the effectiveness of a new medication to treat a deadly virus. The null and alternate hypotheses would be:

H_0: Mean survival rate with new treatment <= Mean survival rate without new treatment

H_1: Mean survival rate with new treatment > Mean survival rate without new treatment

2. CHOOSE THE LEVEL OF SIGNIFICANCE

The level of significance, represented as α, is the probability of rejecting a null hypothesis when it is actually true. To be considered statistically significant, levels of significance range from .1 to .01. Choosing a level of significance of .05 would mean you have only a 5% chance of rejecting a null hypothesis that is true, you are, therefore, 95% confident that you will not reject a null hypothesis that is true. Rejecting a null hypothesis when it is actually true is defined as a Type I error. The level of significance is the chance of making a Type I error.

A type II error, represented by β, occurs when you fail to reject a null hypothesis that is actually false.

3. DETERMINE THE TEST STATISTIC

Up to this point we have discussed the *z* statistic and the *t* statistic. When you are working with means, determining whether to use the *z* or *t* is dependent on the standard deviation. If you know the population standard deviation, you will use the *z* distribution. If not, you will use the *t* distribution. You will also use the *z* statistic when you are dealing with proportions.

You will learn additional test statistics in upcoming chapters. In the third step of the hypothesis testing procedure, you will choose which statistic to compute to make the decision to reject or not reject the null hypothesis.

4. LOCATE THE CRITICAL VALUE (DECISION RULE)

You will use the critical value to determine whether you do reject or do not reject the null hypothesis. The critical value is the test statistic value, *z* or *t* for now, that corresponds to the level of significance, α, you selected in step #2.

The critical value depends on whether you are doing a one-tail or a two-tail test, which you can easily determine by examining the null hypothesis.

One-tail tests examine the data in only one direction and have either <= or >= in the null hypothesis statement. Two-tail tests look at two directions and have = in the null hypothesis statement.

One-tail test examples:

H_o: Mean chips contents >= 8.7 ounces

H_o: Mean survival rate with new treatment <= Mean survival rate without new treatment

Two-tail test example:
H_o: Sandwich assembly time new process = Sandwich assembly time old process

We will delve into a complete example soon. In the meantime, here is a useful tip: if you are using a z statistic, with a two-tail test and a .05 level of significance, the critical z value is 1.96.

5. TAKE A SAMPLE, COMPUTE THE TEST STATISTIC AND MAKE A DECISION

In this final step of the hypothesis testing procedure, we will compute the test statistic. The formula depends on the test statistic and what you are comparing; one sample, two samples, a proportion, or something else.

After computing the test statistic value from your sample, compare it to the critical value you located in the previous step. The comparison of these two values determines whether you do or do not reject the null hypothesis.

You will reject the null hypothesis if the computed value falls within the rejection area, which will be in the tail or tails of the curve. If your computed has an absolute value greater than the critical value, you will reject the null. If the value you compute is smaller, do not reject the null hypothesis.

Using our example two-tail null hypothesis and tip above:

H_0: Sandwich assembly time new process = Sandwich assembly time old process.

Let's say during the five-step hypothesis testing procedure, we set the level of significance at .05 and selected a z-test.

Since this is a two-tail test, 1.96 is the critical value for z. Figure 31 shows the reject and do not reject areas based on a critical z-value value of 1.96.

Do you reject the null hypothesis if the computed z-value is:
    a) 2.5? b) 1.0? c) -1.5? d) -2.5?
        Answer: a) Yes; b) No; c) No; d) Yes.

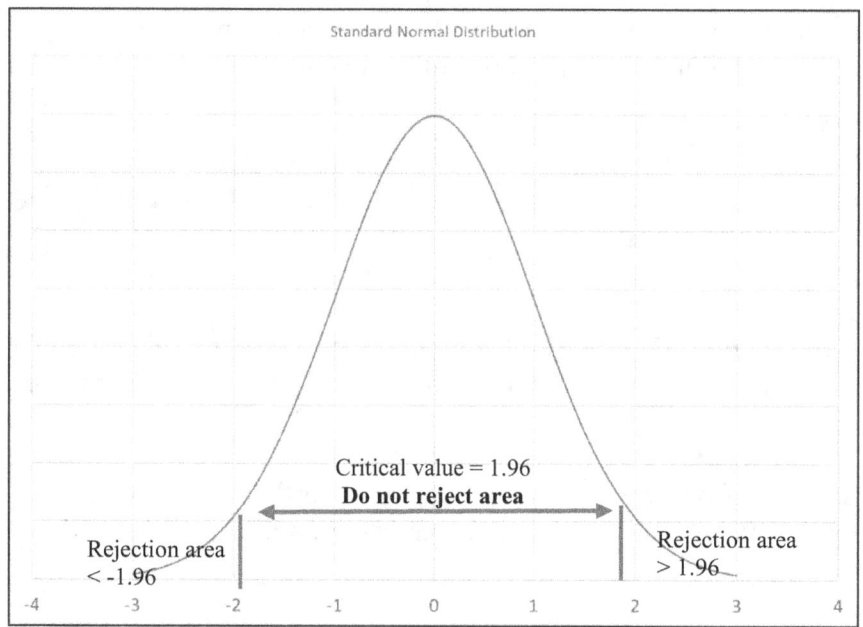

Figure 31: Z-score critical value of 1.96

This process will make sense as we go through an example or two.

It also may be a good time to review Chapter 7 to ensure you understand z and t distribution curves and how to locate z and t scores before you delve into these examples.

One Sample Hypothesis Test – Known Population Standard Deviation

Dan's Deli makes great sandwiches and is experiencing rapid growth. However, customers often complain about the wait after ordering before they receive their sandwiches.

Dan knows that right now it takes an average (mean) of 54 seconds to assemble a sandwich. In Dan's past research, he learned the standard deviation of sandwich production is .7 seconds.

Marisol, one of Dan's employees, has suggested some changes in the process including: rearranging some vegetable locations and changing the order of other ingredients. Marisol is sure her ideas will save time without sacrificing quality. Dan is quite skeptical but agrees to test her method to see if it saves time. Dan will test the method during the assembly of ten sandwiches and use the five-step hypothesis testing process to see if Marisol's method is better or worse than the existing method.

Step 1: State the null and alternate hypothesis:
H_0: $\mu = \bar{x}$ (The sandwich creation is the same, equal, for both methods.)

H_1: $\mu <> \bar{x}$ (The sandwich creation time is not the same, not equal, for both methods.)

Step 2: Select the level of significance:

Dan selects a .1 level of significance. This gives him a 10% chance of making a Type I error, rejecting the null hypothesis when it is actually true.

Step 3: Determine the test statistic:

Since Dan has a population standard deviation from previous research on sandwich assembly, he will use the z statistic.

Step 4: Determine the critical value (decision rule):

This is a two-tailed test because Dan is pretty sure the assembly time will be the same, (i.e. equal) for both methods. However, the alternate hypothesis tests the time as either better or worse, (i.e. not equal).

The critical z-value is determined from the following table in Figure 32. The level of significance determines the "do not reject" and "reject areas" under the normal curve. For two-tail, the area in each tail is 0.1 / 2 = .05 in each tail.

The "do not reject area" is 1 - .1 = .9 for both sides or .45 for each side in the area between the mean and each tail. The area .45 is not in the table. The closest areas are for z scores that equal 1.64 and 1.65. The area .45 is halfway between the listed areas of .4495 and .4505, so our critical value is 1.645, halfway between the two listed z scores.

From this, the takeaway is:
The critical value is: 1.645
.1 level of significance
.05 in each tail
.45 area between mean and tail.

Area under the normal curve										
z	0.00	0.01	0.02	0.03	0.04	0.05	0.06	0.07	0.08	0.09
0	0.0000	0.0040	0.0080	0.0120	0.0160	0.0199	0.0239	0.0279	0.0319	0.0359
0.1	0.0398	0.0438	0.0478	0.0517	0.0557	0.0596	0.0636	0.0675	0.0714	0.0753
0.2	0.0793	0.0832	0.0871	0.0910	0.0948	0.0987	0.1026	0.1064	0.1103	0.1141
0.3	0.1179	0.1217	0.1255	0.1293	0.1331	0.1368	0.1406	0.1443	0.1480	0.1517
0.4	0.1554	0.1591	0.1628	0.1664	0.1700	0.1736	0.1772	0.1808	0.1844	0.1879
0.5	0.1915	0.1950	0.1985	0.2019	0.2054	0.2088	0.2123	0.2157	0.2190	0.2224
0.6	0.2257	0.2291	0.2324	0.2357	0.2389	0.2422	0.2454	0.2486	0.2517	0.2549
0.7	0.2580	0.2611	0.2642	0.2673	0.2704	0.2734	0.2764	0.2794	0.2823	0.2852
0.8	0.2881	0.2910	0.2939	0.2967	0.2995	0.3023	0.3051	0.3078	0.3106	0.3133
0.9	0.3159	0.3186	0.3212	0.3238	0.3264	0.3289	0.3315	0.3340	0.3365	0.3389
1	0.3413	0.3438	0.3461	0.3485	0.3508	0.3531	0.3554	0.3577	0.3599	0.3621
1.1	0.3643	0.3665	0.3686	0.3708	0.3729	0.3749	0.3770	0.3790	0.3810	0.3830
1.2	0.3849	0.3869	0.3888	0.3907	0.3925	0.3944	0.3962	0.3980	0.3997	0.4015
1.3	0.4032	0.4049	0.4066	0.4082	0.4099	0.4115	0.4131	0.4147	0.4162	0.4177
1.4	0.4192	0.4207	0.4222	0.4236	0.4251	0.4265	0.4279	0.4292	0.4306	0.4319
1.5	0.4332	0.4345	0.4357	0.4370	0.4382	0.4394	0.4406	0.4418	0.4429	0.4441
1.6	0.4452	0.4463	0.4474	0.4484	0.4495	0.4505	0.4515	0.4525	0.4535	0.4545
1.7	0.4554	0.4564	0.4573	0.4582	0.4591	0.4599	0.4608	0.4616	0.4625	0.4633
1.8	0.4641	0.4649	0.4656	0.4664	0.4671	0.4678	0.4686	0.4693	0.4699	0.4706
1.9	0.4713	0.4719	0.4726	0.4732	0.4738	0.4744	0.4750	0.4756	0.4761	0.4767
2	0.4772	0.4778	0.4783	0.4788	0.4793	0.4798	0.4803	0.4808	0.4812	0.4817
2.1	0.4821	0.4826	0.4830	0.4834	0.4838	0.4842	0.4846	0.4850	0.4854	0.4857
2.2	0.4861	0.4864	0.4868	0.4871	0.4875	0.4878	0.4881	0.4884	0.4887	0.4890
2.3	0.4893	0.4896	0.4898	0.4901	0.4904	0.4906	0.4909	0.4911	0.4913	0.4916
2.4	0.4918	0.4920	0.4922	0.4925	0.4927	0.4929	0.4931	0.4932	0.4934	0.4936
2.5	0.4938	0.4940	0.4941	0.4943	0.4945	0.4946	0.4948	0.4949	0.4951	0.4952
2.6	0.4953	0.4955	0.4956	0.4957	0.4959	0.4960	0.4961	0.4962	0.4963	0.4964
2.7	0.4965	0.4966	0.4967	0.4968	0.4969	0.4970	0.4971	0.4972	0.4973	0.4974
2.8	0.4974	0.4975	0.4976	0.4977	0.4977	0.4978	0.4979	0.4979	0.4980	0.4981
2.9	0.4981	0.4982	0.4982	0.4983	0.4984	0.4984	0.4985	0.4985	0.4986	0.4986
3	0.4987	0.4987	0.4987	0.4988	0.4988	0.4989	0.4989	0.4989	0.4990	0.4990

Figure 32: Normal distribution table, critical value 1.645

Step 5: Take a sample, compute the test statistic and make a decision

The mean assembly time for the sample of ten sandwiches was 53 seconds using Marisol's revised process.

The formula to compute a z score from one sample mean is:

$$z = \frac{\bar{x} - \mu}{\sigma/\sqrt{n}} \quad \text{So,} \quad z = \frac{53 - 54}{.7/\sqrt{10}} = -4.512$$

Because the computed *z* score, -4.512 falls outside -1.645 and 1.645 and is the **reject area**. Dan is forced to **reject the null hypothesis**. He concludes, at the .1 level of significance, there is a statistically significant difference between the old and Marisol's proposed assembly methods. Based on the test results, Dan promotes Marisol and encourages her to look for even more ways to increase efficiency.

One Sample Hypothesis Test – Unknown Population Standard Deviation

Cindy, an astute high school senior, is considering a nearby private college which advertises is generosity in awarding scholarships and grants. On its website, it states students graduate with and average student loan debt of only $5,000. Cindy believes this number is too low and devises a plan to test her hypothesis. She surveys 15 college seniors a month from graduation.

Step 1: State the null and alternate hypothesis:

H_0: $\mu <= \$5,000$ (The average student loan debt is less than or equal to $5,000.)
H_1: $\mu > \$5,000$ (The average student loan debt is greater than $5,000.)

Step 2: Select the level of significance:

Cindy selects a .05 level of significance. This gives her a 5% chance of rejecting the null hypothesis when it is actually true, a Type I error.

Step 3: Determine the test statistic:

The college only posted the mean, not the standard deviation. Since Cindy does not have the population standard deviation, she will use the *t* statistic.

Step 4: Determine the critical value (decision rule):

This is a one-tail test because Cindy is only checking one direction. To find the critical value, the *t* statistic uses degrees of freedom which, with only one sample, equals one less than sample size, *n – 1*. With a sample size of 15, the degrees of freedom are 14.

As illustrated in Figure 33, using the .05 level of significance for one tail column with 14 degrees of freedom, the critical *t* statistic is 1.761.

	Partial t distribution table					
	One tail					
	0.1	0.05	0.025	0.01	0.005	0.0005
	Two tail					
DF	0.2	0.1	0.05	0.02	0.01	0.001
1	3.078	6.314	12.706	31.821	63.657	636.619
2	1.886	2.920	4.303	6.965	9.925	31.599
3	1.638	2.353	3.182	4.541	5.841	12.924
4	1.533	2.132	2.776	3.747	4.604	8.610
5	1.476	2.015	2.571	3.365	4.032	6.869
6	1.440	1.943	2.447	3.143	3.707	5.959
7	1.415	1.895	2.365	2.998	3.499	5.408
8	1.397	1.860	2.306	2.896	3.355	5.041
9	1.383	1.833	2.262	2.821	3.250	4.781
10	1.372	1.812	2.228	2.764	3.169	4.587
11	1.363	1.796	2.201	2.718	3.106	4.437
12	1.356	1.782	2.179	2.681	3.055	4.318
13	1.350	1.771	2.160	2.650	3.012	4.221
14	1.345	**1.761**	2.145	2.624	2.977	4.140
15	1.341	1.753	2.131	2.602	2.947	4.073

Figure 33: Critical *t* score of 1.761

Step 5: Take a sample, compute the test statistic and make a decision:

Cindy's survey found the following:

Student	Debt	Student	Debt
1	$ 5,500	9	$ -
2	$ 6,000	10	$ 8,800
3	$ 8,500	11	$ 10,000
4	$ 3,000	12	$ 8,000
5	$ 11,000	13	$ 3,000
6	$ 4,200	14	$ 1,080
7	$ 6,600	15	$ 2,550
8	$ 3,000		
$\bar{x}$	$ 5,415.33	S	3347.93511

The formula to compute a *t* score from one sample mean is:

$$t = \frac{\bar{X} - \mu}{s/\sqrt{n}} \quad \text{So,} \quad t = \frac{5415.33 - 5000}{3347.95/\sqrt{15}} = .48046$$

Because the computed *t* score, .48046, is less than 1.791, Cindy does **not reject the null hypothesis**. She cannot conclude, at the .05 level of significance, that the mean student loan debt is greater than $5,000.

Note: Cindy computed a sample mean that is higher than the population mean reported by the college. Why can she not reject the null hypothesis and conclude the school is guilty of false advertising?

Because the .05 level of significance only tolerates a 5% chance that *an actually true* null hypotheses is rejected in error. The bottom line is that Cindy's chances of making a Type I error are greater than what the testing procedure she established will accept. Her chances of making a Type I error are far greater than she specified when she began the test. In fact, the probability (p value) of making a type I error with a t statistic of .48046 and 14 degrees of freedom is 34.47%.

Excel's T.Dist() function computes the *p* value from a *t* statistic.

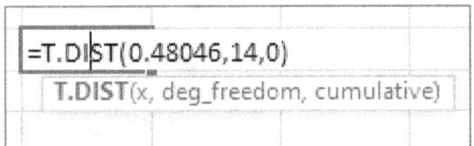

What does Cindy need to do to prove that the college is guilty of false advertising? She needs to increase her sample size (i.e., the number of students in her survey)! This makes sense because a survey of fifteen students is really too small to accuse the college of false advertising.

Hypothesis Test – One Proportion

A recent study that found that 33% of college students change majors in the first three years of college. Mario does not believe that the percentage is this high at the college he attends. He devises a plan to test his hypothesis, which involves randomly surveying fifty seniors at the college he attends.

Step 1: State the null and alternate hypothesis:

    H_0 The proportion of seniors who changed majors is >= 33/100
    H_1 The proportion of seniors who changed majors is < 33/100

Step 2: Select the level of significance:
    Mario selects a .05 level of significance which gives him a 5% chance of making a Type I error.

Step 3: Determine the test statistic:
    Because Mario is working with proportions, he will use the z statistic.

Step 4: Determine the critical value (decision rule):
This is a one-tail test because Mario is only checking one direction, < 33%. The *z* score for an area of .45 (5% in the single tail) equals 1.645 because .45 is halfway between the two areas for the *z* scores of 1.64 and 1.65. Because the rejection area, the tail, is on the left side of the curve, the critical value is actually -1.645 as shown in Figure 34. Remember Mario thinks the percentage is less than 33%.

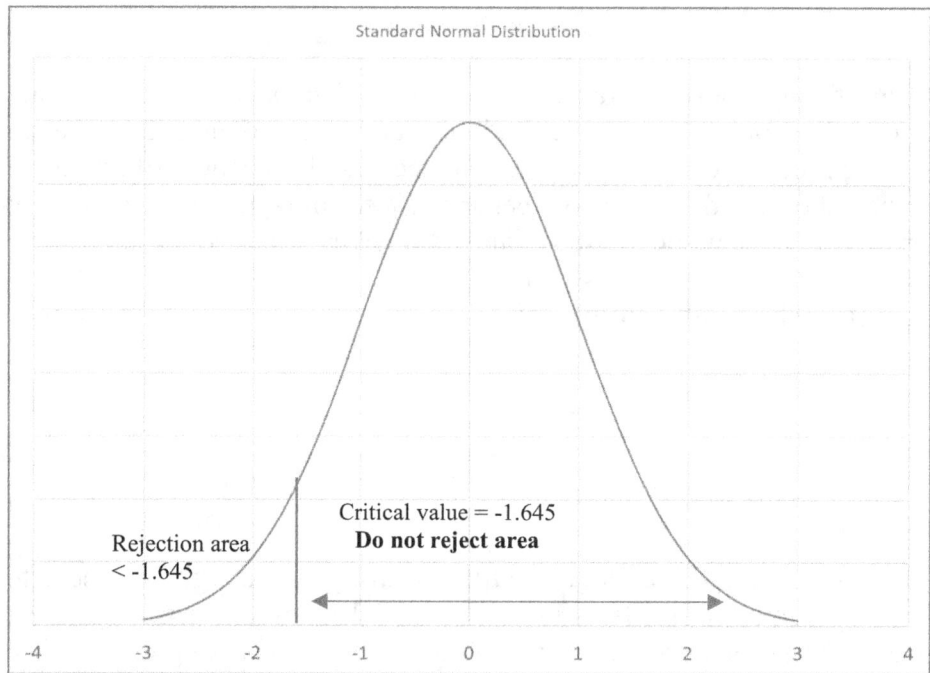

Figure 34: Critical Value -1.645

Area under the normal curve										
z	0.00	0.01	0.02	0.03	0.04	0.05	0.06	0.07	0.08	0.09
0	0.0000	0.0040	0.0080	0.0120	0.0160	0.0199	0.0239	0.0279	0.0319	0.0359
0.1	0.0398	0.0438	0.0478	0.0517	0.0557	0.0596	0.0636	0.0675	0.0714	0.0753
0.2	0.0793	0.0832	0.0871	0.0910	0.0948	0.0987	0.1026	0.1064	0.1103	0.1141
0.3	0.1179	0.1217	0.1255	0.1293	0.1331	0.1368	0.1406	0.1443	0.1480	0.1517
0.4	0.1554	0.1591	0.1628	0.1664	0.1700	0.1736	0.1772	0.1808	0.1844	0.1879
0.5	0.1915	0.1950	0.1985	0.2019	0.2054	0.2088	0.2123	0.2157	0.2190	0.2224
0.6	0.2257	0.2291	0.2324	0.2357	0.2389	0.2422	0.2454	0.2486	0.2517	0.2549
0.7	0.2580	0.2611	0.2642	0.2673	0.2704	0.2734	0.2764	0.2794	0.2823	0.2852
0.8	0.2881	0.2910	0.2939	0.2967	0.2995	0.3023	0.3051	0.3078	0.3106	0.3133
0.9	0.3159	0.3186	0.3212	0.3238	0.3264	0.3289	0.3315	0.3340	0.3365	0.3389
1	0.3413	0.3438	0.3461	0.3485	0.3508	0.3531	0.3554	0.3577	0.3599	0.3621
1.1	0.3643	0.3665	0.3686	0.3708	0.3729	0.3749	0.3770	0.3790	0.3810	0.3830
1.2	0.3849	0.3869	0.3888	0.3907	0.3925	0.3944	0.3962	0.3980	0.3997	0.4015
1.3	0.4032	0.4049	0.4066	0.4082	0.4099	0.4115	0.4131	0.4147	0.4162	0.4177
1.4	0.4192	0.4207	0.4222	0.4236	0.4251	0.4265	0.4279	0.4292	0.4306	0.4319
1.5	0.4332	0.4345	0.4357	0.4370	0.4382	0.4394	0.4406	0.4418	0.4429	0.4441
1.6	0.4452	0.4463	0.4474	0.4484	0.4495	0.4505	0.4515	0.4525	0.4535	0.4545
1.7	0.4554	0.4564	0.4573	0.4582	0.4591	0.4599	0.4608	0.4616	0.4625	0.4633
1.8	0.4641	0.4649	0.4656	0.4664	0.4671	0.4678	0.4686	0.4693	0.4699	0.4706
1.9	0.4713	0.4719	0.4726	0.4732	0.4738	0.4744	0.4750	0.4756	0.4761	0.4767
2	0.4772	0.4778	0.4783	0.4788	0.4793	0.4798	0.4803	0.4808	0.4812	0.4817
2.1	0.4821	0.4826	0.4830	0.4834	0.4838	0.4842	0.4846	0.4850	0.4854	0.4857
2.2	0.4861	0.4864	0.4868	0.4871	0.4875	0.4878	0.4881	0.4884	0.4887	0.4890
2.3	0.4893	0.4896	0.4898	0.4901	0.4904	0.4906	0.4909	0.4911	0.4913	0.4916
2.4	0.4918	0.4920	0.4922	0.4925	0.4927	0.4929	0.4931	0.4932	0.4934	0.4936
2.5	0.4938	0.4940	0.4941	0.4943	0.4945	0.4946	0.4948	0.4949	0.4951	0.4952
2.6	0.4953	0.4955	0.4956	0.4957	0.4959	0.4960	0.4961	0.4962	0.4963	0.4964
2.7	0.4965	0.4966	0.4967	0.4968	0.4969	0.4970	0.4971	0.4972	0.4973	0.4974
2.8	0.4974	0.4975	0.4976	0.4977	0.4977	0.4978	0.4979	0.4979	0.4980	0.4981
2.9	0.4981	0.4982	0.4982	0.4983	0.4984	0.4984	0.4985	0.4985	0.4986	0.4986
3	0.4987	0.4987	0.4987	0.4988	0.4988	0.4989	0.4989	0.4989	0.4990	0.4990

Step 5: Take a sample, compute the test statistic and make a decision:

Of the fifty seniors Mario surveyed, ten reported changing majors in their first three years.

The formula to compute a z score from one proportion is:

$$z = \frac{p - p^*}{\sqrt{\frac{p^*(1-p^*)}{n}}} \quad \text{So,} \quad z = \frac{.2 - .33}{\sqrt{\frac{.33 * (1 - .33)}{50}}} = -1.955$$

p^* = population proportion, p = sample proportion

Mario computed a z score which is more negative than the critical value. He will reject the null hypothesis and conclude, that at least at his school, the reported proportion may not be correct.

Surviving Statistics

Chapter Highlights

- **Five step hypothesis testing procedure**
  1. State the null and alternate hypothesis.
  2. Choose the level of significance.
  3. Determine the test statistic.
  4. Locate the critical value (decision rule).
  5. Take a sample, compute the test statistic and make a decision.

- **One sample hypothesis test – known population standard deviation**

$$z = \frac{\bar{X} - \mu}{\sigma/\sqrt{n}}$$

- **One sample hypothesis test – unknown population standard deviation**

$$t = \frac{\bar{X} - \mu}{s/\sqrt{n}}$$

- **Hypothesis test – one proportion**

$$z = \frac{p - p^*}{\sqrt{\frac{p^*(1-p^*)}{n}}}$$

Chapter 11: Two-Sample Hypothesis Testing

This chapter will cover:

Two Samples Known Population Standard Deviation

Two Samples Unknown Population Standard Deviation

Two Proportions

Dependent and Independent Samples

Chapter 11: Two sample hypothesis testing

In the previous chapter, we explored hypothesis testing comparing one sample's mean or proportion to a population mean or proportion. In this chapter, we will discuss the computations needed in hypothesis testing to compare two samples.

Two Sample Hypothesis Testing – Known Population Standard Deviation

We will compare two sample means, each representing a different population. We will, however, need the standard deviation for each population. To help understand this procedure, here is an example.

Keisha, a new administrator for Quad-States Hospital, is interested in the satisfaction ratings for day and evening shift nurses. Patients are asked to rate their nurses on several items on a scale of 1 to 5. As she began her research, she was told that both shifts have equal patient satisfaction ratings and both shifts have a standard deviation of .2.

From listening to a few patients since arriving here, Keisha believes the night shift may be doing a better job caring for patients. She sets out to test this hypothesis.

Step 1: State the null and alternate hypothesis:

H_0 Night shift satisfaction ratings <= day shift ratings
H_1 Night shift satisfaction ratings > day shift ratings

Step 2: Select the level of significance:
Keisha selects a .02 level of significance. This gives her a 2% chance of making a Type I error.

Step 3: Determine the test statistic:
Because Keisha has the historic population standard deviation, she will use the z distribution.

Step 4: Determine the critical value (decision rule):
This is a one-tail test because Keisha is only checking one direction. The z score that equates to an area of .48 (2% in the single tail) and using the table in Figure 35, this is close to 2.055.

Area under the normal curve										
z	0.00	0.01	0.02	0.03	0.04	0.05	0.06	0.07	0.08	0.09
0	0.0000	0.0040	0.0080	0.0120	0.0160	0.0199	0.0239	0.0279	0.0319	0.0359
0.1	0.0398	0.0438	0.0478	0.0517	0.0557	0.0596	0.0636	0.0675	0.0714	0.0753
0.2	0.0793	0.0832	0.0871	0.0910	0.0948	0.0987	0.1026	0.1064	0.1103	0.1141
0.3	0.1179	0.1217	0.1255	0.1293	0.1331	0.1368	0.1406	0.1443	0.1480	0.1517
0.4	0.1554	0.1591	0.1628	0.1664	0.1700	0.1736	0.1772	0.1808	0.1844	0.1879
0.5	0.1915	0.1950	0.1985	0.2019	0.2054	0.2088	0.2123	0.2157	0.2190	0.2224
0.6	0.2257	0.2291	0.2324	0.2357	0.2389	0.2422	0.2454	0.2486	0.2517	0.2549
0.7	0.2580	0.2611	0.2642	0.2673	0.2704	0.2734	0.2764	0.2794	0.2823	0.2852
0.8	0.2881	0.2910	0.2939	0.2967	0.2995	0.3023	0.3051	0.3078	0.3106	0.3133
0.9	0.3159	0.3186	0.3212	0.3238	0.3264	0.3289	0.3315	0.3340	0.3365	0.3389
1	0.3413	0.3438	0.3461	0.3485	0.3508	0.3531	0.3554	0.3577	0.3599	0.3621
1.1	0.3643	0.3665	0.3686	0.3708	0.3729	0.3749	0.3770	0.3790	0.3810	0.3830
1.2	0.3849	0.3869	0.3888	0.3907	0.3925	0.3944	0.3962	0.3980	0.3997	0.4015
1.3	0.4032	0.4049	0.4066	0.4082	0.4099	0.4115	0.4131	0.4147	0.4162	0.4177
1.4	0.4192	0.4207	0.4222	0.4236	0.4251	0.4265	0.4279	0.4292	0.4306	0.4319
1.5	0.4332	0.4345	0.4357	0.4370	0.4382	0.4394	0.4406	0.4418	0.4429	0.4441
1.6	0.4452	0.4463	0.4474	0.4484	0.4495	0.4505	0.4515	0.4525	0.4535	0.4545
1.7	0.4554	0.4564	0.4573	0.4582	0.4591	0.4599	0.4608	0.4616	0.4625	0.4633
1.8	0.4641	0.4649	0.4656	0.4664	0.4671	0.4678	0.4686	0.4693	0.4699	0.4706
1.9	0.4713	0.4719	0.4726	0.4732	0.4738	0.4744	0.4750	0.4756	0.4761	0.4767
2	0.4772	0.4778	0.4783	0.4788	0.4793	0.4798	0.4803	0.4808	0.4812	0.4817
2.1	0.4821	0.4826	0.4830	0.4834	0.4838	0.4842	0.4846	0.4850	0.4854	0.4857
2.2	0.4861	0.4864	0.4868	0.4871	0.4875	0.4878	0.4881	0.4884	0.4887	0.4890
2.3	0.4893	0.4896	0.4898	0.4901	0.4904	0.4906	0.4909	0.4911	0.4913	0.4916
2.4	0.4918	0.4920	0.4922	0.4925	0.4927	0.4929	0.4931	0.4932	0.4934	0.4936
2.5	0.4938	0.4940	0.4941	0.4943	0.4945	0.4946	0.4948	0.4949	0.4951	0.4952
2.6	0.4953	0.4955	0.4956	0.4957	0.4959	0.4960	0.4961	0.4962	0.4963	0.4964
2.7	0.4965	0.4966	0.4967	0.4968	0.4969	0.4970	0.4971	0.4972	0.4973	0.4974
2.8	0.4974	0.4975	0.4976	0.4977	0.4977	0.4978	0.4979	0.4979	0.4980	0.4981
2.9	0.4981	0.4982	0.4982	0.4983	0.4984	0.4984	0.4985	0.4985	0.4986	0.4986
3	0.4987	0.4987	0.4987	0.4988	0.4988	0.4989	0.4989	0.4989	0.4990	0.4990

Figure 35: Critical Z score of 2.055

Step 5: Take a sample, compute the test statistic and make a decision
Keisha randomly surveyed 25 patients and asked them to rate their day shift nurse. She then surveyed another 30 patients and asked them to rate their night shift nurses. The mean for the day shift nurses was 4.55 and the night shift nurses was 4.7.

The formula to compute a z score from two samples is:

$$z = \frac{\bar{x}_1 - \bar{x}_2}{\sqrt{\frac{\sigma_1^2}{n_1} + \frac{\sigma_2^2}{n_2}}} \quad \text{So,} \quad z = \frac{4.70 - 4.55}{\sqrt{\frac{.04}{30} + \frac{.04}{25}}} = 2.77$$

Keisha's computed a *z* score is larger than the critical value. She will reject the null hypothesis and conclude, at least for this sample, night shift nurses have higher ratings than day shift nurses.

Two Sample– Unknown Standard Deviation

Many times, as we have already discussed, when you are doing your own research you will not have the population standard deviation available. In these cases, you will likely use the *t* distribution.

To illustrate a two-sample hypothesis test using the *t* distribution we will revisit Keisha and her hypothesis on nursing ratings. This time though she either does not have the population standard deviation or does not trust the one she was provided.

Step 1: State the null and alternate hypothesis:

H_0 Night shift satisfaction ratings <= day shift ratings
H_1 Night shift satisfaction ratings > day shift ratings

Step 2: Select the level of significance:
Keisha selects a .05 level of significance. This gives her a 5% chance of making a Type I error.

Step 3: Determine the test statistic:
Because Keisha does not have the population standard deviation, she will use the *t* statistic.

Step 4: Determine the critical value (decision rule):
This is a one-tail test because Keisha is only checking one direction. The t distribution requires Keisha to use degrees for freedom as well as the level of significance to locate the critical value. With two samples, the degrees of freedom are: n1 + n2 – 2, or 53 in this case.

Using a table or Excel, Keisha locates a critical t value of 1.674.

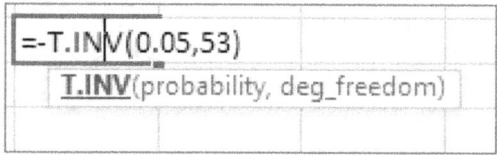

If you examine the Excel formula closely you will notice the (-) at the beginning. Because this function returns the left tail value, which is negative, placing the minus sign at the beginning causes the t critical value to be positive.

Step 5: Take a sample, compute the test statistic and make a decision
Keisha took two samples, 25 day patients and 30 evening patients. The results with the mean and sample standard deviation are as follows:

	A	B	C
1	Patient	Day	Night
2	1	4.6	4.6
3	2	4.8	4.8
4	3	4.8	4.8
5	4	5	5
6	5	3.2	3.2
7	6	3.8	3.8
8	7	4.8	4.8
9	8	5	5
10	9	3.9	3.9
11	10	3.75	4.8
12	11	4.8	4.8
13	12	5	5
14	13	4.6	4.7
15	14	3.8	4.5
16	15	5	5
17	16	4.4	4.6
18	17	4.6	4.8
19	18	4.7	4.8
20	19	5	5
21	20	4.7	4.8
22	21	4.7	4.8
23	22	5	5
24	23	4.6	4.6
25	24	4.8	4.8
26	25	4.8	4.8
27	26		4.8
28	27		5
29	28		4.8
30	29		4.8
31	30		5
32	Mean	4.566	4.70333
33	Stdev	0.48707	0.39955

Computing a *t* score from two samples is somewhat complex if you do it manually. Excel has an excellent tool that we will discuss later. For now though, we will go over the manual method.

The first step in computing a two-sample *t* test score is to compute the pooled variance. The formula for doing this is:

$$s_p^2 = \frac{(n_1 - 1)s_1^2 + (n_2 - 1)s_2^2}{n_1 + n_2 - 2}$$

After computing the pooled variance, you use it in the formula to compute a t score, which is:

$$t = \frac{\bar{X}_1 - \bar{X}_2}{\sqrt{s_p^2 \left(\frac{1}{n_1} + \frac{1}{n_2}\right)}}$$

So, for Keisha, the pooled variance becomes:

$$s_p^2 = \frac{(30-1).15964 + (25-1).23723}{30 + 25 - 2}$$

The pooled variance = .19478

Notice that we are using the night shift, the sample with the highest mean as sample 1.

Now that we have the pooled variance, we can compute the *t* statistic as follows:

$$t = \frac{4.7033 - 4.566}{\sqrt{.19478 \left(\frac{1}{30} + \frac{1}{25}\right)}} = 1.1488$$

In this case, the computed *t* score is less than the critical value, so Keisha will not reject the null hypothesis. She cannot conclude that night nurses receive higher satisfaction scores than day shift nurses do.

When Keisha conducted the first hypothesis test using the *z* statistic, she was able to reject the null hypothesis. However, using the *t* test she is not able to do so. This is because the standard deviation she computed with her sample is much larger than the assumed population standard deviation (.02) she used with the z test.

Try it in Excel:

Excel's data analysis add-in has a *t* test option that computes the *t* statistic after selecting the data from both samples. From the data analysis tools, choose *t* –Test Two sample assuming equal variances.

> *Note: Even though the variances of the day and night shift samples appear to differ in this example, they are not unequal from a statistical standpoint. In the next chapter, you will learn to conduct a hypothesis test to examine the equality of variance.*

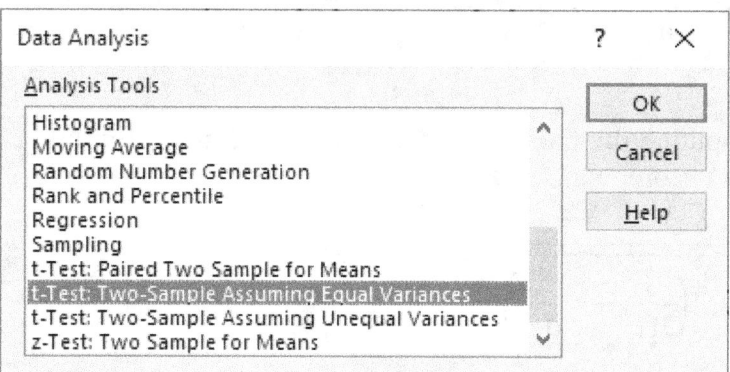

Next, select the two ranges containing the data from each sample. Select the desired alpha level, the level of significance, and click OK.

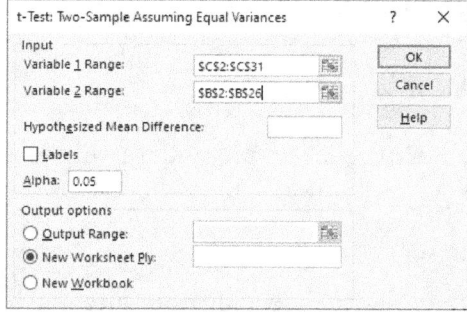

The result will display t statistics for both one and two tail tests. Excel also displays the critical values for each test in the output.

t-Test: Two-Sample Assuming Equal Variances		
	Variable 1	Variable 2
Mean	4.703333333	4.566
Variance	0.159643678	0.237233
Observations	30	25
Pooled Variance	0.194778616	
Hypothesized Mean Difference	0	
df	53	
t Stat	1.149091495	
P(T<=t) one-tail	0.127838822	
t Critical one-tail	1.674116237	
P(T<=t) two-tail	0.255677643	
t Critical two-tail	2.005745995	

Two Sample Hypothesis Test – Proportions

Keisha is continuing her efforts to analyze and improve patient satisfaction. While patients interact with several nurses, Keisha realizes they seem to remember the nurse who discharged them, their last contact with the hospital, more than all the others. Keisha attempts to see if there is a difference in the overall patient satisfaction ratings between patients discharged by day shift nurses and those discharged by night shift nurses. Over the next week, Keisha randomly called several discharged patients and asked them if they were satisfied with the services offered at the hospital, allowing only "yes" or "no" as valid answers. Keisha uses patient records to see if they were discharged by day or night shift nurses.

Step 1: State the null and alternate hypothesis:

H_0 Day shift nurse patient satisfaction = Night shift nurse patient satisfaction
H_1 Day shift nurse patient satisfaction <> Night shift nurse patient satisfaction

Step 2: Select the level of significance:
Keisha selects a .05 level of significance. This gives her a 5% chance of making a type I error.

Step 3: Determine the test statistic:
Because Keisha is working with proportions, she will use the z distribution.

Step 4: Determine the critical value (decision rule):
This is a two-tail test because Keisha is checking both directions, "different than". With a .05 level of significance, each tail will have an area of 0.025. Finding the area of 0.475 in the normal distribution table, Keisha determines the critical value is 1.96.

Step 5: Take a sample, compute the test statistic and make a decision
Keisha surveys 55 discharged patients and finds that 25 were discharged by day shift nurses and 30 by night shift nurses. The results were as follows:
Day shift discharge patients satisfied: 20/25
Night shift discharge patients satisfied: 25/30

The formula for computing a z score comparing two proportions has two steps.

The first step is to compute the pooled proportion.

The formula to do this is:

$$P_c = \frac{x_1 + x_2}{n_1 + n_2} = \frac{20 + 25}{25 + 30} = .8182$$

Surviving Statistics

After computing the pooled proportion, you can then compute the z score with this formula:

$$z = \frac{p_1 - p_2}{\sqrt{\frac{p_c(1-p_c)}{n_1} + \frac{p_c(1-p_c)}{n_2}}} = \frac{.8 - .8333}{\sqrt{\frac{.8182(1-.8182)}{25} + \frac{.8182(1-.8182)}{30}}}$$

The computed Z score is =.31884, falls within the "Do not reject" area because it is not greater than 1.96 or less than -1.96. Keisha will not reject the null hypothesis and conclude, based on this limited survey, the satisfaction ratings between patients discharged by day and evening nurses are equal.

Two Sample Hypothesis Test – Dependent Samples

The samples discussed and used in the examples so far are independent samples. That is, the patients surveyed about day nurses were not the same patients surveyed about night nurses. Other than being in the same hospital, there is no direct relationship between the two samples.

Conversely, a dependent sample is the same sample surveyed twice. The second survey is usually after an intervention such as a training, an adjustment, or a medication. The hypothesis revolves around the intervention and tests if the intervention resulted in a change, positive or negative. The same individuals are surveyed before and after the intervention.

To illustrate a hypothesis test using dependent samples, assume Keisha decides to see if a training program can improve individual nurses' patient satisfaction ratings. She decides to measure their mean ratings before and after the training. Keisha randomly selects ten nurses and records their patient satisfaction ratings for one week. After they attend customer service training, she records the same nurses' ratings for the week after they complete training. Keisha is hopeful the training will improve patient satisfaction and sets out to test her hypothesis.

Step 1: State the null and alternate hypothesis:

    H₀   Ratings after training <= Ratings before training
    H₁   Ratings after training > Ratings before training

Step 2: Select the level of significance:
Keisha selects a .05 level of significance. This gives her a 5% chance of making a Type I error.

Step 3: Determine the test statistic:
We will assume Keisha does not have the population standard deviation and therefore, she uses the *t* distribution.

Step 4: Determine the critical value (decision rule):

This is a one-tail test because Keisha is checking only one direction, "greater than". The sample size is 10, even though the nurses were sampled twice. That means the degrees of freedom are 9. The critical value from the *t* distribution table in Figure 36 is 1.833.

		Partial *t* distribution table				
			One tail			
	0.1	0.05	0.025	0.01	0.005	0.0005
			Two tail			
DF	0.2	0.1	0.05	0.02	0.01	0.001
1	3.078	6.314	12.706	31.821	63.657	636.619
2	1.886	2.920	4.303	6.965	9.925	31.599
3	1.638	2.353	3.182	4.541	5.841	12.924
4	1.533	2.132	2.776	3.747	4.604	8.610
5	1.476	2.015	2.571	3.365	4.032	6.869
6	1.440	1.943	2.447	3.143	3.707	5.959
7	1.415	1.895	2.365	2.998	3.499	5.408
8	1.397	1.860	2.306	2.896	3.355	5.041
9	1.383	1.833	2.262	2.821	3.250	4.781
10	1.372	1.812	2.228	2.764	3.169	4.587
11	1.363	1.796	2.201	2.718	3.106	4.437
12	1.356	1.782	2.179	2.681	3.055	4.318
13	1.350	1.771	2.160	2.650	3.012	4.221
14	1.345	1.761	2.145	2.624	2.977	4.140
15	1.341	1.753	2.131	2.602	2.947	4.073

Figure 36: Partial t distribution table

Step 5: Take a sample, compute the test statistic and make a decision

Keisha gathers the ratings for the ten nurses before and after the training as shown below:

	Rating	
Nurse	Before	After
Adams	3.90	4.40
Brown	4.80	4.90
Green	3.70	4.10
Lee	4.00	4.80
Martinez	4.70	4.50
Park	4.80	4.40
Smith	3.70	4.20
Smith	3.90	4.40
Tanner	4.50	4.40
Washington	4.30	4.80

The formula to compute a t statistic from dependent (paired) samples is:

$$t = \frac{\bar{d}}{s_d/\sqrt{n}}$$

In this formula s_d is the standard deviation of the differences between the before and after values. It is computed as:

$$s_d = \sqrt{\frac{\Sigma(d - \bar{d})^2}{n - 1}}$$

	A	B	C	D	E
1		\multicolumn{2}{c}{Rating}		
2	Nurse	Before	After	Difference (d)	$(d - \bar{d})^2$
3	Adams	3.90	4.40	0.50	0.0576
4	Brown	4.80	4.90	0.10	0.0256
5	Green	3.70	4.10	0.40	0.0196
6	Lee	4.00	4.80	0.80	0.2916
7	Martinez	4.70	4.50	-0.20	0.2116
8	Park	4.80	4.40	-0.40	0.4356
9	Smith	3.70	4.20	0.50	0.0576
10	Smith	3.90	4.40	0.50	0.0576
11	Tanner	4.50	4.40	-0.10	0.1296
12	Washington	4.30	4.80	0.50	0.0576
13	Sum	42.30		2.60	1.344
14			Mean ($\bar{d}$)	0.26	
15			Standard deviation of d	0.38644	

Note that we compute the difference by subtracting the mean rating before then training from the mean after the training.

After computing the standard deviation of the differences, the t statistic is computed as:

$$\frac{0.26}{.38644/\sqrt{10}} = 2.128$$

The computed t stat is larger than the critical value of 1.833, so Keisha rejects the null hypothesis and concludes the training resulted in higher ratings for the nurses.

Try it in Excel:

Excel's data analysis add-in has a paired t test option that computes a t statistic for dependent samples.

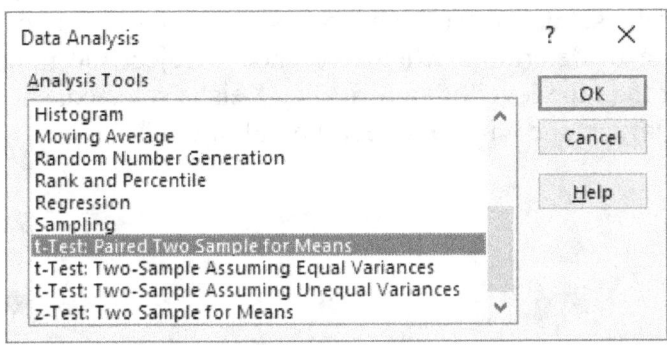

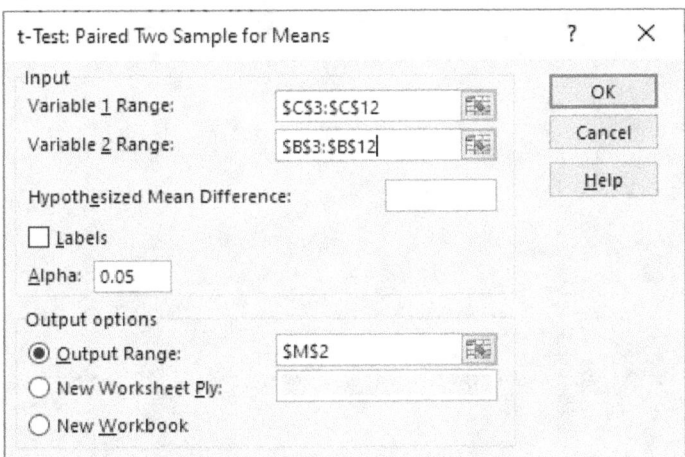

To avoid computing a negative t statistic, the "after training" sample is *Variable range 1*.

M	N	O
t-Test: Paired Two Sample for Means		
	Variable 1	Variable 2
Mean	4.49	4.23
Variance	0.06989	0.19789
Observations	10	10
Pearson Correlation	0.50358	
Hypothesized Mean Difference	0	
df	9	
t Stat	2.12762	
P(T<=t) one-tail	0.03113	
t Critical one-tail	1.83311	
P(T<=t) two-tail	0.06226	
t Critical two-tail	2.26216	

As with the other *t* test options in Excel's data analysis feature, the output table displays critical values for both one and two tail test as well as *p* values.

The *p* value for this one tail test is .03113. Keisha set up to test the hypothesis with a .05 level of significance. Keisha can also use the *p* value to determine whether or not to reject the null hypothesis. If the computed *p* value is smaller than the level of significance Keisha selected, she can reject the null hypothesis. If the *p* value is larger, she will not reject the null hypothesis.

Chapter Highlights

All hypothesis testing will use the same five steps introduced in the last chapter. However, you will use different formula depending on what you are testing.

- **Two sample hypothesis testing – known population standard deviation**

$$Z = \frac{\bar{X}_1 - \bar{X}_2}{\sqrt{\frac{\sigma_1^2}{n_1} + \frac{\sigma_2^2}{n_2}}}$$

- **Two sample– unknown standard deviation**

  - **Pooled variance:**

  $$s_p^2 = \frac{(n_1-1)s_1^2 + (n_2-1)s_2^2}{n_1 + n_2 - 2}$$

  - **Computing the t score:**

  $$t = \frac{\bar{X}_1 - \bar{X}_2}{\sqrt{s_p^2 \left(\frac{1}{n_1} + \frac{1}{n_2}\right)}}$$

- **Two sample hypothesis test – proportions**

  - **Pooled proportion:**

  $$P_c = \frac{x_1 + x_2}{n_1 + n_2}$$

  - **Z score for two proportions:**

  $$z = \frac{p_1 - p_2}{\sqrt{\frac{p_c(1-p_c)}{n_1} + \frac{p_c(1-p_c)}{n_2}}}$$

- **Two Sample hypothesis test – dependent samples**

  - **T score from dependent (paired) samples:**

  $$t = \frac{\bar{d}}{s_d/\sqrt{n}}$$

  - **Standard deviation of the differences between the before and after the intervention:**

  $$s_d = \sqrt{\frac{\Sigma(d - \bar{d})}{n - 1}}$$

Surviving Statistics

Chapter 12: ANOVA (Analysis of Variance)

This chapter will cover:

Analysis of Variance – Two Populations

Constructing an ANOVA Table: Three or More Means

Chapter 12: ANOVA (Analysis of Variance)

We use the ANOVA to compare the variance of two populations or to compare the means of more than two samples. We accomplish these two goals differently, but both will use a test statistic we have not yet used, the F statistic

Here are some attributes of the F distribution you should know:
- The F distribution, like the T distribution uses degrees of freedom.
- The F statistic will never equal a negative value, but it can be zero.

We will start this chapter with a comparison of two population variances and a simple formula to compute the F statistic. Then we will create a complete ANOVA table as we compare the means from three samples.

Comparing two population variances

Although it does not appear on every patient satisfaction survey, Keisha keeps running into one consistent complaint: the discharge process takes too long. Patients comment that the waiting time after the doctor releases them to when they can actually leave the hospital is too long. Keisha wants to compare the dispersion (variance) in times between patients released by their doctors before noon and those released after noon. She decided to compare the variances of these two populations. To do this, she surveyed a sample of 12 patients released in the morning and 9 patients released after noon. The results are as follows:

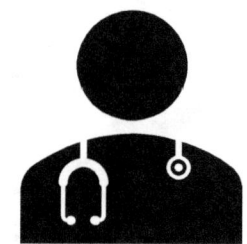

	A	B	C
1		Time until discharge	
2		Morning	Afternoon
3		60	50
4		45	20
5		80	22
6		39	39
7		22	22
8		40	31
9		70	28
10		65	50
11		55	33
12		40	
13		66	
14		29	
15	Mean	50.9167	32.777778
16	Variance	315.174	131.69444
17	Standard Deviation	17.7531	11.47582

Step 1: State the null and alternate hypothesis:

    H_0 Variance of morning releases = Variance of afternoon releases
    H_1 Variance of morning releases <> Variance of afternoon releases

Step 2: Select the level of significance:
Keisha selects a .02 level of significance. This gives her a 2% chance of making a Type I error.

Step 3: Determine the test statistic:
Because Keisha is comparing the variances of these two populations, she will use the F distribution.

Step 4: Determine the critical value (decision rule):
This is a two-tail test because Keisha is checking both directions, which means Keisha is looking for .01 in each tail. Even though the F distribution does not support negative values, we will still treat this as a two-tail test.

The F distribution uses two degrees of freedom, one for the numerator and one for the denominator. The numerator will be the sample size of the sample with the largest variance, and the denominator is the size of the sample with the smallest variance.

The degrees of freedom for the numerator and the denominator are $n - 1$, so 11 for the numerator and 8 for the denominator. Locating the .01 table and the degrees of freedom for the numerator and denominator, the critical F value is 5.734.

F distribution critical values — 0.01 Level of signifiance
Degrees of freedom

Numerator					Denominator							
	1	2	3	4	5	6	7	8	9	10	11	12
1	4052	5000	5403	5625	5764	5859	5928	5981	6022	6056	6083	6106
2	98.503	99.000	99.166	99.249	99.299	99.333	99.356	99.374	99.388	99.399	99.408	99.416
3	34.116	30.817	29.457	28.710	28.237	27.911	27.672	27.489	27.345	27.229	27.133	27.052
4	21.198	18.000	16.694	15.977	15.522	15.207	14.976	14.799	14.659	14.546	14.452	14.374
5	16.258	13.274	12.060	11.392	10.967	10.672	10.456	10.289	10.158	10.051	9.963	9.888
6	13.745	10.925	9.780	9.148	8.746	8.466	8.260	8.102	7.976	7.874	7.790	7.718
7	12.246	9.547	8.451	7.847	7.460	7.191	6.993	6.840	6.719	6.620	6.538	6.469
8	11.259	8.649	7.591	7.006	6.632	6.371	6.178	6.029	5.911	5.814	5.734	5.667
9	10.561	8.022	6.992	6.422	6.057	5.802	5.613	5.467	5.351	5.257	5.178	5.111
10	10.044	7.559	6.552	5.994	5.636	5.386	5.200	5.057	4.942	4.849	4.772	4.706
11	9.646	7.206	6.217	5.668	5.316	5.069	4.886	4.744	4.632	4.539	4.462	4.397
12	9.330	6.927	5.953	5.412	5.064	4.821	4.640	4.499	4.388	4.296	4.220	4.155
13	9.074	6.701	5.739	5.205	4.862	4.620	4.441	4.302	4.191	4.100	4.025	3.960
14	8.862	6.515	5.564	5.035	4.695	4.456	4.278	4.140	4.030	3.939	3.864	3.800
15	8.683	6.359	5.417	4.893	4.556	4.318	4.142	4.004	3.895	3.805	3.730	3.666
16	8.531	6.226	5.292	4.773	4.437	4.202	4.026	3.890	3.780	3.691	3.616	3.553
17	8.400	6.112	5.185	4.669	4.336	4.102	3.927	3.791	3.682	3.593	3.519	3.455
18	8.285	6.013	5.092	4.579	4.248	4.015	3.841	3.705	3.597	3.508	3.434	3.371
19	8.185	5.926	5.010	4.500	4.171	3.939	3.765	3.631	3.523	3.434	3.360	3.297
20	8.096	5.849	4.938	4.431	4.103	3.871	3.699	3.564	3.457	3.368	3.294	3.231

Step 5: Take a sample, compute the test statistic and make a decision

Keisha has already taken the sample, so the next step is to compute the F statistic. The formula to do this is:

$$F = \frac{s_1^2}{s_2^2}$$

Keisha computes an F of 2.393. Since this F statistic is smaller than the critical value, Keisha does not reject the null hypothesis. Even though the variances appear to differ, she cannot conclude that is actually the case at the .02 significance level.

If Keisha is uneasy with these results, she could conduct another study with a larger sample size. Notice the critical values in the F table decrease substantially as the degrees of freedom increase. With a larger sample size, Keisha's decision could be different.

Completing an ANOVA table – three or more means

The t test works well for comparing two sample means. However, it becomes very tedious to use the t statistic to compare three or more means. The ANOVA is a better approach. The final result of the ANOVA will be an F statistic that can be used to reject or not reject the null hypothesis. When we are comparing three or more means the null hypothesis is that the means are equal. The alternate is that at least one mean differs from the others. (Theoretically, we are testing if the means of the populations that the samples were taken from are equal.)

Software programs, like Excel and specialized statistical applications, make creating an ANOVA table quick and easy. However, we will step through the manual creation process first to help you better understand the values that the ANOVA table displays. We will also use ANOVA tables in upcoming chapters.

Before we begin the five-step hypothesis testing process, here is an illustration of an ANOVA table and the values we will be computing. This table will make more sense as we step through the process.

ANOVA Table				
Sources of variation	**Sum of Squares**	**Degrees of freedom**	**Mean Square**	**F**
Treatments	SST	$k - 1$	SST / $(k - 1)$ = MST	*MST/MSE*
Error	SSE	$n - k$	SSE / $(n - k)$ = MSE	
Total	SS Total	$n - 1$		

Keisha is still interested in the time patients wait after being released by their doctor until they are actually discharged. She wants to compare the average (mean) waiting time for those released by their doctors in the morning, afternoon, and weekends. She wants to know if there is a statistically significant difference in the mean waiting times for these three patient populations.

To test her hypothesis, Keisha collects the following data:

Time until discharge		
Morning	Afternoon	Weekend
60	50	80
45	20	90
80	22	45
39	39	60
22	22	50
40	31	40
70	28	45
65	50	55
55	33	45
40		60
66		
29		

Step 1: State the null and alternate hypothesis:

    H_0 Mean morning = mean afternoon = mean weekend
          Or $\mu_1 = \mu_2 = \mu_3$

    H_1 All three means are not equal

Step 2: Select the level of significance:
Keisha will again select a .02 level of significance, giving her a 2% chance of making a Type I error.

Step 3: Determine the test statistic:
Keisha will construct an ANOVA table to compare the means of these three populations with the F distribution.

Step 4: Determine the critical value (decision rule):
This is a two-tail test because Keisha is checking both directions. This means we are looking for .01 in each tail. Even though the F distribution does not support negative values, we will still treat this as a two-tail test.

The degrees of freedom are determined as follows:
    Numerator $= k - 1 = 3 - 1 = 2$
        k is the number of categories or populations, 3 in this case.

    Denominator are $n - k = 31 - 3 = 28$
        n is the total of all three samples which is this case is 31.

The degrees of freedom for the numerator are 2 and the degrees of freedom for the denominator are 28. Locating the .01 table and the degrees of freedom for the numerator and denominator, the critical F value is 5.453.

Surviving Statistics

F distribution critical values — 0.01 Level of significance
Degrees of freedom

Numerator	\ Denominator 1	2	3	4	5	6	7	8	9	10	11	12
10	10.044	7.559	6.552	5.994	5.636	5.386	5.200	5.057	4.942	4.849	4.772	4.706
11	9.646	7.206	6.217	5.668	5.316	5.069	4.886	4.744	4.632	4.539	4.462	4.397
12	9.330	6.927	5.953	5.412	5.064	4.821	4.640	4.499	4.388	4.296	4.220	4.155
13	9.074	6.701	5.739	5.205	4.862	4.620	4.441	4.302	4.191	4.100	4.025	3.960
14	8.862	6.515	5.564	5.035	4.695	4.456	4.278	4.140	4.030	3.939	3.864	3.800
15	8.683	6.359	5.417	4.893	4.556	4.318	4.142	4.004	3.895	3.805	3.730	3.666
16	8.531	6.226	5.292	4.773	4.437	4.202	4.026	3.890	3.780	3.691	3.616	3.553
17	8.400	6.112	5.185	4.669	4.336	4.102	3.927	3.791	3.682	3.593	3.519	3.455
18	8.285	6.013	5.092	4.579	4.248	4.015	3.841	3.705	3.597	3.508	3.434	3.371
19	8.185	5.926	5.010	4.500	4.171	3.939	3.765	3.631	3.523	3.434	3.360	3.297
20	8.096	5.849	4.938	4.431	4.103	3.871	3.699	3.564	3.457	3.368	3.294	3.231
21	8.017	5.780	4.874	4.369	4.042	3.812	3.640	3.506	3.398	3.310	3.236	3.173
22	7.945	5.719	4.817	4.313	3.988	3.758	3.587	3.453	3.346	3.258	3.184	3.121
23	7.881	5.664	4.765	4.264	3.939	3.710	3.539	3.406	3.299	3.211	3.137	3.074
24	7.823	5.614	4.718	4.218	3.895	3.667	3.496	3.363	3.256	3.168	3.094	3.032
25	7.770	5.568	4.675	4.177	3.855	3.627	3.457	3.324	3.217	3.129	3.056	2.993
26	7.721	5.526	4.637	4.140	3.818	3.591	3.421	3.288	3.182	3.094	3.021	2.958
27	7.677	5.488	4.601	4.106	3.785	3.558	3.388	3.256	3.149	3.062	2.988	2.926
28	7.636	5.453	4.568	4.074	3.754	3.528	3.358	3.226	3.120	3.032	2.959	2.896
29	7.598	5.420	4.538	4.045	3.725	3.499	3.330	3.198	3.092	3.005	2.931	2.868

Step 5: Take a sample, compute the test statistic and make a decision

Keisha has already taken the sample, so the next step is to construct the ANOVA table and compute the F statistic.

The first step in creating the ANOVA is to create the grand mean, the mean of all the samples. For these samples, the grand mean is 47.613 (1476/31).

	Time until discharge			
	Morning	Afternoon	Weekend	
	60	50	80	
	45	20	90	
	80	22	45	
	39	39	60	
	22	22	50	
	40	31	40	
	70	28	45	
	65	50	55	
	55	33	45	
	40		60	
	66			
	29			
Column total	611	295	570	1476
n	12	9	10	31
Mean	50.91667	32.777778	57	47.6129

ANOVA Table				
Sources of variation	Sum of Squares	Degrees of freedom	Mean Square	F
Treatments	SST	k – 1	SST / (k – 1) = MST	MST/MSE
Error	SSE	n – k	SSE / (n – k) = MSE	
Total	SS Total	n – 1		

The next step is to compute the Sum of Squares Total (SS Total).

This is computed as:

$$SS\ Total = \sum (X - \bar{X}_G)^2$$

where $\bar{x}_G$ is the grand mean.

To compute *SS Total*, we subtract the grand mean from each value, square it and them sum all the squared differences for each sample.

	Morning	$(X - \bar{X}_G)^2$	Afternoon	$(X - \bar{X}_G)^2$	Weekends	$(X - \bar{X}_G)^2$
	60	153.44017	50	5.698231	80	1048.924
	45	6.8272633	20	762.4724	90	1796.666
	80	1048.924	22	656.0208	45	6.827263
	39	74.182102	39	74.1821	60	153.4402
	22	656.02081	22	656.0208	50	5.698231
	40	57.956296	31	275.9886	40	57.9563
	70	501.1821	28	384.666	45	6.827263
	65	302.31113	50	5.698231	55	54.5692
	55	54.569199	33	213.5369	45	6.827263
	40	57.956296			60	153.4402
	66	338.08533				
	29	346.44017				
Total		3597.8949		3034.284		3291.176
SS Total		**9923.3548**				

Completing the ANOVA table as we go, we can now fill it in as follows:

ANOVA Table				
Sources of variation	Sum of Squares	Degrees of freedom	Mean Square	F
Treatments	SST	2	SST / (k – 1) = MST	MST/MSE
Error	SSE	28	SSE / (n – k) = MSE	
Total	9923.355	30		

The next step is to create the Sum of Squares Error.

$$\sum (X - \bar{X}_c)^2$$

Surviving Statistics

We compute this by taking each value in the sample and subtracting the treatment (sample) mean from that value. We then square the difference for each value and sum all the errors for each treatment (sample). Finally, as the term implies, SSE is the summed treatment (sample) errors.

SSE Computation						
	Morning	$(X - \bar{X}_1)^2$	Afternoon	$(X - \bar{X}_2)^2$	Weekends	$(X - \bar{X}_3)^2$
	60	82.506944	50	296.6049	80	529
	45	35.006944	20	163.2716	90	1089
	80	845.84028	22	116.1605	45	144
	39	142.00694	39	38.71605	60	9
	22	836.17361	22	116.1605	50	49
	40	119.17361	31	3.160494	40	289
	70	364.17361	28	22.82716	45	144
	65	198.34028	50	296.6049	55	4
	55	16.673611	33	0.049383	45	144
	40	119.17361			60	9
	66	227.50694				
	29	480.34028				
Mean	50.91667		32.777778		57	
Total		3466.9167		1053.556		2410
SSE	6930.472					

Once we have these two key values, SS Total and SSE, we can construct the rest of the ANOVA table with minimal calculations.

Having SS Total and SSE, we can easily compute SST with: $SST = SS\ Total - SSE$

All the other computations are based on SST and SSE. So, the completed table, including the F statistic is as follows:

ANOVA Table				
Sources of variation	Sum of Squares	Degrees of freedom	Mean Square	F
Treatments	9923.355-6930.472 =2992.883	2	SST / (k − 1) = MST =2992.883 / 2 = 1496.442	MST/MSE =1496.442/247.517 = 6.046
Error	6930.472	28	SSE / (n − k) = MSE =6930.472 / (31 − 3) = 247.517	
Total	9923.355	30		

After constructing the ANOVA table and computing an F statistic of 6.046, Keisha can reject the null hypothesis because the F statistic she computed is larger than the critical F value of 5.734

All three of the means are not equal at the .02 level of significance. Essentially, Keisha can be at least 98% sure that the wait times in the morning, afternoon, and weekends are not the same.

Try it in Excel:

Excel's data analysis add-in has the ANOVA single factor option that will create an ANOVA table and compute the F statistic for more than two samples.

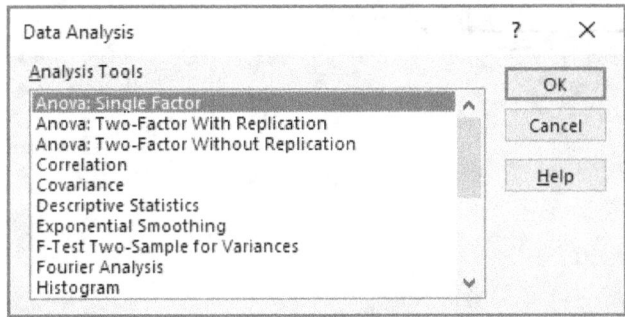

Be sure to set the correct Alpha level, which will be the level of significance / 2 because you are really doing a two tailed test.

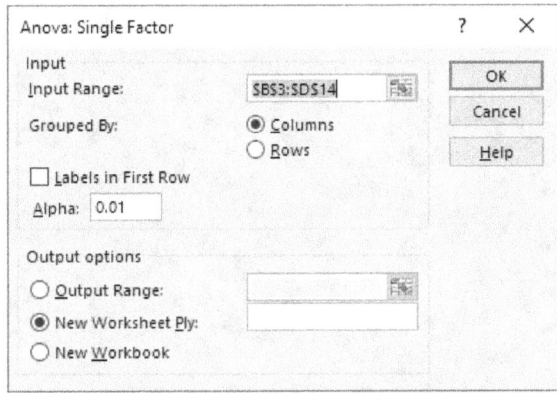

You should be able to understand these values after constructing an ANOVA manually. Notice that Excel names SST, "Between Groups", and SSE "Within Groups".

Anova: Single Factor

SUMMARY

Groups	Count	Sum	Average	Variance
Column 1	12	611	50.91667	315.1742
Column 2	9	295	32.77778	131.6944
Column 3	10	570	57	267.7778

ANOVA

Source of Variation	SS	df	MS	F	P-value	F crit
Between Groups	2992.883	2	1496.441	6.045816	0.006568	5.452937
Within Groups	6930.472	28	247.5169			
Total	9923.355	30				

Chapter Highlights

We use the ANOVA to compare the variance of two populations or to compare the means of more than two samples. The ANOVA uses the F distribution,

- **Comparing two population variances** $$F = \frac{s_1^2}{s_2^2}$$

- **Completing an ANOVA table – three or more means**

The easiest way to complete an ANOVA table is to compute SSE and SS Total. All the other values can be derived from these two computations.

ANOVA Table

Sources of variation	Sum of Squares	Degrees of freedom	Mean Square	F
Treatments	SST	$k-1$	SST / $(k-1)$ = MST	MST/MSE
Error	SSE	$n-k$	SSE / $(n-k)$ = MSE	
Total	SS Total	$n-1$		

$$SS\ Total = \sum (X - \bar{X}_G)^2$$

SSE: $\sum (X - \bar{X}_c)^2$

Computing SSE is rather complex. Refer to the chapter for the steps.

Surviving Statistics

Chapter 13: Correlation and Linear Regression

This chapter will cover:
- **Correlation Coefficient**
- **Regression Equation**
- **Coefficient of Determination**
- **Multiple Regression Analysis**

Chapter 13: Correlation and Linear Regression

You may recall the scatterplot we discussed in Chapter 4 as shown in Figure 37. The purpose of a scatterplot is to display a possible relationship or correlation between two variables. The scatterplot shown below displays the relationship between the amount of lemonade Larry sold and the day's high temperature. We also briefly introduced you to the concept of a numerical value that represents the same thing, the correlation coefficient.

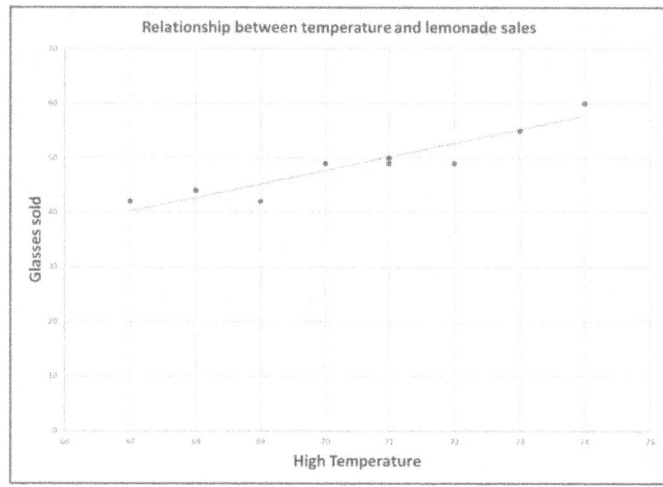

In this chapter, we will compute the correlation coefficient that indicates the relationship strength between two variables. We will use linear regression to attempt to determine the relationships between variables and predict values for one variable based on the values of one or more other variables.

Figure 37: A sample scatterplot

Dependent and Independent Variables

Both correlation and regression analysis deal with dependent and independent variables. Independent variables are usually designated as X and dependent variables are designated with Y. In the lemonade stand example, the scatterplot seemed to indicate that lemonade sales were affected by the temperature. With linear regression, we are attempting to predict values of the dependent variable, Y, based on values of the dependent variable, X.

Correlation Coefficient

The correlation coefficient measures the strength of the linear relationship between two variables. Its value can be negative or positive and ranges from -1 to +1 as shown in Figure 38. In the lemonade stand example, the scatter plot suggests a strong positive correlation between the high temperature and the amount of lemonade sold.

	Correlation Coefficient Strength	
Perfect negative correlation	No Correlation	Perfect positive correlation

Strong Negative Correlation	Moderate Negative Correlation	Weak Negative Correlation		Weak Positive Correlation	Moderate Positive Correlation	Strong Positive Correlation	
-1.00	-0.75	-0.50	-0.25	0.00	0.25	0.50	0.75	1.00

Figure 38: Interpreting the correlation coefficient

The formula to compute a correlation coefficient is:

$$r = \frac{\sum(X - \bar{X})(Y - \bar{Y})}{(n-1)s_x s_y}$$

The computations for Larry's lemonade are as follows:

Day	Lemonade Sales Temperature -X	# glasses sold - Y	$(X - \bar{X})$	$(Y - \bar{Y})$	$(Y - \bar{Y})(Y - \bar{Y})$
Sunday	72	49	1.143	-0.857	-0.980
Monday	71	50	0.143	0.143	0.020
Tuesday	68	44	-2.857	-5.857	16.735
Wednesday	71	49	0.143	-0.857	-0.122
Thursday	73	55	2.143	5.143	11.020
Friday	67	42	-3.857	-7.857	30.306
Saturday	74	60	3.143	10.143	31.878
Sunday	70	49	-0.857	-0.857	0.735
Monday	71	50	0.143	0.143	0.020
Tuesday	68	44	-2.857	-5.857	16.735
Wednesday	71	49	0.143	-0.857	-0.122
Thursday	73	55	2.143	5.143	11.020
Friday	69	42	-1.857	-7.857	14.592
Saturday	74	60	3.143	10.143	31.878
Mean	70.857	49.857			
Standard Deviation	2.248	5.908			
Total					163.714
				r =	0.948

The computed correlation coefficient of .948 confirms our observation from the scatterplot. There is a very strong positive relationship between the day's high temperature and lemonade sales.

Regression Analysis

The result of regression analysis is an equation that we can use to predict a value for a dependent variable based on independent variable(s). Linear regression rests on the least squares principle, which creates a line minimizing the sum of the squares between the actual and predicted values of Y, the dependent variable.

The linear equation will have an intercept and a slope. After computing the slope and intercept you can then use various values of X, the independent variable, to predict value for Y.

The linear regression formula is:

$$\hat{Y} = a + bX$$

where $\hat{Y}$ is the predicted value of Y.

SLOPE

The slope determines how much Y increases with each increase in X. The formula to determine the slope is:

$$b = r \frac{S_y}{S_x}$$

We have already computed the correlation coefficient, r for the lemonade sales example. We have also previously computed the standard deviation for both variables. So, we can compute the slope as:

$$b = .948 \frac{5.908}{2.248} = 2.49145$$

INTERCEPT

The Y intercept, the value of Y when X = zero is computed as:

$$a = \bar{Y} - b\bar{X}$$

As with the slope, we have everything we need to compute the intercept as:

$$a = 49.857 - 2.49145(70.857) = -126.68$$

REGRESSION FORMULA

Now that we have computed the slope and intercept, we can complete the regression formula as:

$$\hat{Y} = -126.68 + 2.49145(X)$$

So, if the high temperature of the day is going to be 75°F, we would predict Larry would sell:

Predicted glasses sold = -126.687 + 2.49145(75) = 60.18

Try it in Excel:

Computations for regression analysis can be time consuming to do manually. Statistical software and Excel make them much easier. We'll go over the steps to perform regression in Excel and discuss some important regression analysis parameters which Excel includes in it summary output table.

In Data Analysis, choose Regression.

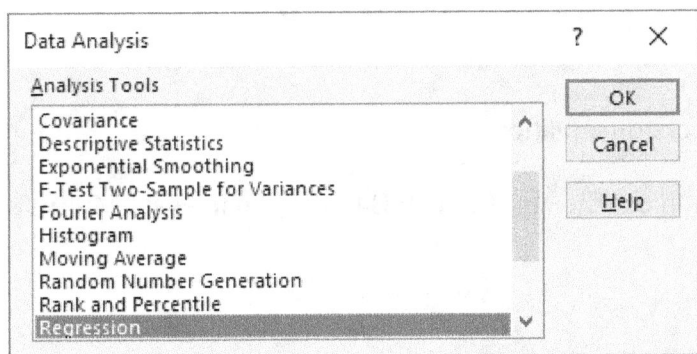

Next, highlight the cells for the Y (dependent variable) and the X (independent variable) and click OK.

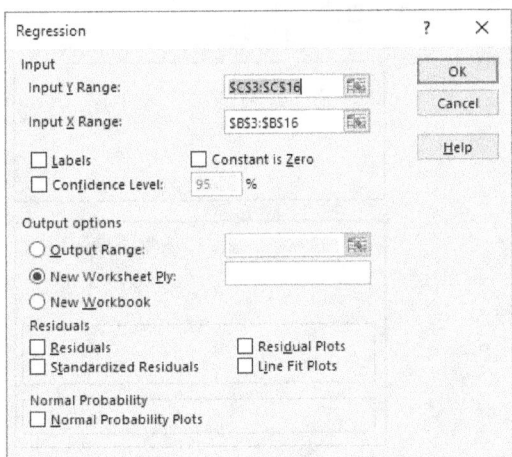

Surviving Statistics

	A	B	C	D	E	F	G	H	I
1	SUMMARY OUTPUT								
2									
3	Regression Statistics								
4	Multiple R	0.948							
5	R Square	0.899							
6	Adjusted R Square	0.891							
7	Standard Error	1.955							
8	Observations	14							
9									
10	ANOVA								
11		df	SS	MS	F	Significance F			
12	Regression	1	407.862	407.862	106.742	0.000			
13	Residual	12	45.852	3.821					
14	Total	13	453.714						
15									
16		Coefficients	Standard Error	t Stat	P-value	Lower 95%	Upper 95%	Lower 95.0%	Upper 95.0%
17	Intercept	-126.670	17.094	-7.410	0.000	-163.914	-89.425	-163.914	-89.425
18	X Variable 1	2.491	0.241	10.332	0.000	1.966	3.017	1.966	3.017

The intercept we computed, -126.670, appears in this output in cell B17.

The slope, 2.491, appears in cell B18.

From these two cells, you can create the regression equation.

Notice that Excel also computed the correlation coefficient, r in cell B4. Excel names it "Multiple R," but it is still the correlation coefficient.

STANDARD ERROR OF THE ESTIMATE

The standard error of the estimate measures how widely the predicted values and actual values are dispersed. The larger the standard error of the estimate, the more the actual data points vary from the created regression line (predicted values). If you use Excel or other statistical software to compute a regression, this value is calculated for you. You can see the standard error in cell B7 of the regression output.

The formula to compute the standard error of the estimate manually is:

$$S_{y*x} = \sqrt{\frac{\sum(Y-\hat{Y})^2}{n-2}}$$

COEFFICIENT OF DETERMINATION

The correlation coefficient, r, measures the strength of the relationship between the dependent and independent variable. The coefficient of determination, r^2, measures the proportion of variance in the dependent variable, Y, that can be explained by the dependent variable, X. The coefficient of determination is computed by simply squaring r. In the regression output, the coefficient of determination shown in cell B5 is .899. This tells us that roughly 90% of the variability in lemonade sales can be explained by the day's high temperature.

Multiple Regression

In Larry's lemonade example, we examined the relationship between sales and one independent variable, the day's high temperature. However, Larry thinks there may be other factors that affect lemonade sales. The purpose of multiple regression is to create a regression equation that more accurately predicts values of the dependent variable because it includes additional statistically significant independent variables.

Continuing on with Larry's Lemonade, Larry believes he sells more lemonade when the humidity is higher. He also believes he sells less lemonade on days on which rain is likely. He also thinks he sells more lemonade on workdays than he does on the weekend.

In creating a multiple regression Larry, or whoever does the analysis, will throw every possible variable into the mix. Then, after the regression is computed, each potential independent variable will be evaluated to see if it is a statistically significant predictor of the dependent variable. Variables that are not statistically significant predictors are eliminated from the regression equation and the result is an equation that can accurately predict the independent variable.

The next figure shows the data Larry has collected to analyze in a multiple regression.

Dummy Variables

You may notice that Workday has a value of 0 or 1. Workday is a qualitative variable, not quantitative. Dummy variables allow some qualitative variables, those that use a nominal scale to be used in regression. In this case workdays are coded as 1 and weekend days as 0.

Day	Lemonade Sales # glasses sold - Y	High Temperature [F]	Noon Humidity	Chance of rain	Workday
Sunday	49	72	35	30	0
Monday	50	71	22	25	1
Tuesday	44	68	50	35	1
Wednesday	49	71	38	29	1
Thursday	55	73	60	10	1
Friday	42	67	55	35	1
Saturday	60	74	49	0	0
Sunday	49	70	55	33	0
Monday	50	71	60	24	1
Tuesday	44	68	48	35	1
Wednesday	49	71	55	25	1
Thursday	55	73	68	10	1
Friday	42	69	51	30	1
Saturday	60	74	68	5	0

Performing a multiple regression is very difficult to do manually. We will illustrate this example using Excel and discuss its output.

When performing a multiple regression in Excel, highlight all the potential dependent variables in the *X* range.

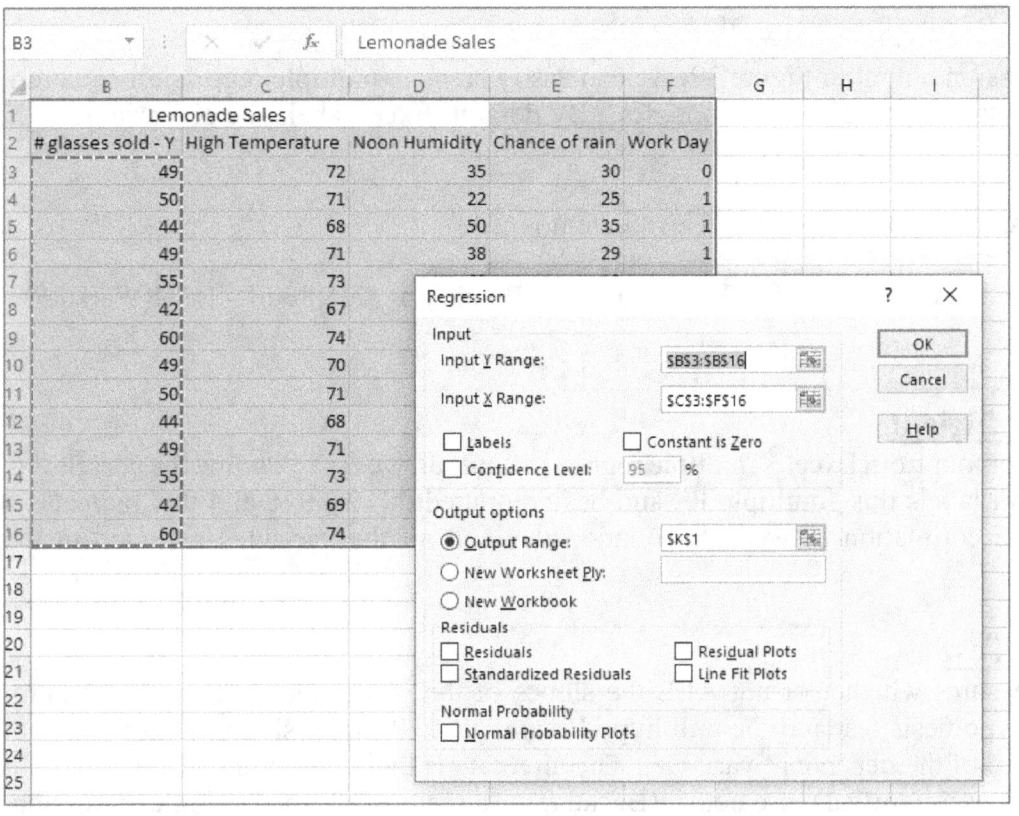

SUMMARY OUTPUT								
Regression Statistics								
Multiple R	0.976							
R Square	0.952							
Adjusted R Square	0.931							
Standard Error	1.548							
Observations	14.000							
ANOVA								
	df	SS	MS	F	Significance F			
Regression	4	432.151	108.038	45.093	0.000			
Residual	9	21.563	2.396					
Total	13	453.714						
	Coefficients	Standard Error	t Stat	P-value	Lower 95%	Upper 95%	Lower 95.0%	Upper 95.0%
Intercept	-29.263	37.888	-0.772	0.460	-114.971	56.446	-114.971	56.446
Temp	1.205	0.492	2.452	0.037	0.093	2.317	0.093	2.317
Humidity	0.007	0.040	0.178	0.863	-0.084	0.098	-0.084	0.098
Rain	-0.235	0.093	-2.532	0.032	-0.445	-0.025	-0.445	-0.025
Workday	-1.654	1.067	-1.550	0.156	-4.069	0.760	-4.069	0.760

Figure 39: Output of multiple regression with Excel

THE MULTIPLE REGRESSION EQUATION

From Excel's regression output in Figure 39, we can easily create a multiple regression equation. The illustration shows the names of the variables. By default, Excel labels the X Variable 1, X Variable 2, and so on.

The multiple regression equation, before we evaluate it is:

Predicted Lemonade Sales = -29.263 + 1.205(Temp) + .007(Humidity) - .235(Rain) – 1.654(Workday)

EVALUATING A MULTIPLE REGRESSION OUTPUT

Correlation Coefficient

As we examine the output from Excel's multiple regression output, we first examine the correlation coefficient, r. Excel labels this "multiple R" and has computed this as .976 and this indicates a very strong positive correlation between lemonade sales and all the variables included in the multiple regression.

The Global (F) Test

The global test measures whether or not ALL the slopes of the independent variables are zero. This is actually a hypothesis test and the null hypothesis is that all of the slopes are zero, or not significant predictors of the dependent variable. The alternate is that at least one is not zero, and therefore, at least one is a significant predictor. (Because we already performed a single regression. We already know that temperature is a significant predictor, so we are actually only concerned about the others.)

To determine the critical value for F with a .05 level of significance, we use 4 degrees of freedom in the numerator, computed as p (number of all Y and X variables) – 1. The degrees of freedom for the denominator are 9, computed as $n – p$. Notice that Excel displays the degrees of freedom in its output.

The critical value from the table in Figure 40 is 4.718. Remember we divide the level of significance by 2 with the F distribution. The F statistic that Excel computed for this example is 45.093. This is far larger than the critical value, so we will reject the null hypothesis and conclude that at least one of the independent variables included in the analysis is a significant predictor. In practice, we don't have to locate the critical F values if we are using Excel. Instead, we can use the p value Excel labels "*Significance F*". The p value is essentially zero in this example, which makes sense when we see how large the computed F is in relation to the critical value at the .05 level of significance.

Because we have rejected the null hypothesis in the Global test, we will continue to evaluate the variables.

F distribution critical values		0.025 Level of significance										
Degrees of freedom												
					Denominator							
Numerator	1	2	3	4	5	6	7	8	9	10	11	12
5	10.007	8.434	7.764	7.388	7.146	6.978	6.853	6.757	6.681	6.619	6.568	6.525
6	8.813	7.260	6.599	6.227	5.988	5.820	5.695	5.600	5.523	5.461	5.410	5.366
7	8.073	6.542	5.890	5.523	5.285	5.119	4.995	4.899	4.823	4.761	4.709	4.666
8	7.571	6.059	5.416	5.053	4.817	4.652	4.529	4.433	4.357	4.295	4.243	4.200
9	7.209	5.715	5.078	4.718	4.484	4.320	4.197	4.102	4.026	3.964	3.912	3.868
10	6.937	5.456	4.826	4.468	4.236	4.072	3.950	3.855	3.779	3.717	3.665	3.621
11	6.724	5.256	4.630	4.275	4.044	3.881	3.759	3.664	3.588	3.526	3.474	3.430
12	6.554	5.096	4.474	4.121	3.891	3.728	3.607	3.512	3.436	3.374	3.321	3.277
13	6.414	4.965	4.347	3.996	3.767	3.604	3.483	3.388	3.312	3.250	3.197	3.153
14	6.298	4.857	4.242	3.892	3.663	3.501	3.380	3.285	3.209	3.147	3.095	3.050
15	6.200	4.765	4.153	3.804	3.576	3.415	3.293	3.199	3.123	3.060	3.008	2.963
16	6.115	4.687	4.077	3.729	3.502	3.341	3.219	3.125	3.049	2.986	2.934	2.889
17	6.042	4.619	4.011	3.665	3.438	3.277	3.156	3.061	2.985	2.922	2.870	2.825
18	5.978	4.560	3.954	3.608	3.382	3.221	3.100	3.005	2.929	2.866	2.814	2.769
19	5.922	4.508	3.903	3.559	3.333	3.172	3.051	2.956	2.880	2.817	2.765	2.720
20	5.871	4.461	3.859	3.515	3.289	3.128	3.007	2.913	2.837	2.774	2.721	2.676
21	5.827	4.420	3.819	3.475	3.250	3.090	2.969	2.874	2.798	2.735	2.682	2.637
22	5.786	4.383	3.783	3.440	3.215	3.055	2.934	2.839	2.763	2.700	2.647	2.602
23	5.750	4.349	3.750	3.408	3.183	3.023	2.902	2.808	2.731	2.668	2.615	2.570
24	5.717	4.319	3.721	3.379	3.155	2.995	2.874	2.779	2.703	2.640	2.586	2.541

Figure 40: F distribution .025 level of significance

Evaluating Each Independent Variable

The Global test lets us know that at least one of the independent variables is a statistically significant predictor of Larry's lemonade sales. The next step is to evaluate each variable to see if it should be included in the regression equation. We can do this by testing the slope of each variable with a *t* test.

The evaluation process is also a hypothesis test. The null is that the slope is zero, which would mean it is not a significant predictor. The alternate is that the slope is not zero and is therefore, a significant predictor of the dependent variable. In this example, we will use the .05 level of significance.

This is a two-tail test. We compute the degrees of freedom as $n - (k + 1)$, where n (number of observations) = 14 and *k* is the number of independent variables. The result is *9* degrees of freedom. We can see that Excel computed this for us and the result is the same as that for the denominator of the global *F* test, $n - p$. The critical *t* value with 9 degrees of freedom is 1.833. You can use the table in Figure 36 on page 129 to verify this value. Knowing the critical value, we can then compare the *t* stat for each independent variable and determine if it is a significant predictor of lemonade sales.

The *t* values Excel computed are as follows:

    Temperature = 2.452
    Humidity   = 0.178
    Rain   = -2.532
    Workday   = -1.550

From the *t* values, we know that only *temperature* and *rain* are significant predictors of lemonade sales. The other variables, while they may play some part in lemonade sales, do not pass the test based on the criteria Larry established, the .05 level of significance.

Rather than going to all the effort of computing the degrees of freedom and looking up the critical *t* value, Larry could have simply used the *p* values that Excel displays in its regression output for each of these variables. In Excel's output only *temperature* and *rain* have *p* values that are <= .05, the level of significance Larry established.

Creating a More Accurate Multiple Regression Equation

Larry threw every possible independent variable he could think of into the multiple regression. Not all the variables he included proved to be statistically significant predictors at the .05 level of significance. The next step is to run another regression, but this time only including the two significant predictors, *temperature* and *rain*.

SUMMARY OUTPUT

Regression Statistics

Multiple R	0.969
R Square	0.939
Adjusted R Square	0.928
Standard Error	1.580
Observations	14

ANOVA

	df	SS	MS	F	Significance F
Regression	2	426.248	213.124	85.354	0.000
Residual	11	27.466	2.497		
Total	13	453.714			

	Coefficients	Standard Error	t Stat	P-value	Lower 95%	Upper 95%	Lower 95.0%	Upper 95.0%
Intercept	-48.598	31.917	-1.523	0.156	-118.848	21.651	-118.848	21.651
Temp	1.461	0.427	3.423	0.006	0.522	2.400	0.522	2.400
Rain	-0.217	0.080	-2.714	0.020	-0.394	-0.041	-0.394	-0.041

In the new regression the global test (*F*), is larger than the previous regression and the *p* value for the *F* test is essentially zero. This lets us know that at least one of the independent variables included in this regression is a significant predictor. (But we already knew that.)

The new regression model more accurately models the data than the first regression which included variables that were not statistically significant predictors of lemonade sales. You should also notice that the t values are larger, and the p values are smaller than the first regression. Now, we can create a multiple regression equation that includes only significant independent variables. The revised multiple regression equation becomes:

Predicted Lemonade Sales = $-48.598 + 1.461$(Temp) $- .217$(Rain)

For a high temperature of 80^0F and a 40% chance of rain,

$-48.598 + 1.461(80) - .217(40) = 62.602$ lemonade sales

Chapter Highlights

- **Dependent and Independent Variables**

Dependent variables depend on the independent variable(s). With linear regression, we are attempting to predict values of the dependent variable. Dependent variables are usually designated Y and independent variables, X.

- **Correlation Coefficient -r**

The correlation coefficient measures the strength of the relationship between two variables. It can range from -1 to +1.

$$r = \frac{\sum(X - \bar{X})(Y - \bar{Y})}{(n-1)s_x s_y}$$

- **Regression Analysis**

Linear regression results in an equation used to predict a value for a dependent variable based on independent variable(s).

$$\hat{Y} = a + bX$$

  o **Slope**

  The slope determines how much Y increases with each increase in X.

  $$b = r\frac{s_y}{s_x}$$

  o **Intercept**

  The value of Y when X is zero:

  $$a = \bar{Y} - b\bar{X}$$

- **Standard Error of the Estimate**

The standard error of the estimate measures how widely the predicted values and actual values are dispersed on the regression line.

$$s_{y*x} = \sqrt{\frac{\sum(Y - \hat{Y})^2}{n-2}}$$

- **Coefficient of Determination – r^2**

The coefficient of determination, r^2, measures the proportion of variance in the dependent variable, Y, that can be explained by the independent variable, X.

- **Dummy Variables**

Dummy variables allow some qualitative variables, those that use a nominal scale to be used in regression.

- **The Multiple Regression Equation**

The multiple regression equation

Predicted Y = a + B$_1$ (X$_1$) + B$_2$ (X$_2$) + B$_3$ (X$_3$)...

- **Evaluating a Multiple Regression Output**

    o **Correlation Coefficient**

    Examine the correlation coefficient, r. to see if there is a relationship strong enough to continue.

    o **The Global (F) Test**

    The global test measures whether or not ALL the slopes of the independent variables are zero.

    o **Evaluating Each Independent Variable**

    The next step is to evaluate each variable to see if it should be included in the regression equation. We do this by testing the slope of each variable with a T test.

    o **Creating a more accurate multiple regression equation**

    Run another regression using only statistically significant independent variables.

Chapter 14: Goodness of Fit Tests

This chapter will cover:
- Goodness of Fit
  - Equal Expected Frequencies
  - Unequal Expected Frequencies

Chapter 14: Goodness of Fit Tests

Goodness of fit tests allow us to do hypothesis testing on data that uses the nominal scale of measurement. As we discussed in the first chapter, we cannot compute means and several other measurements from data that uses the nominal scale. We can, however, measure and analyze frequencies or counts. In this chapter, we will be discussing goodness of fit tests, which compare actual frequencies to the counts we expected. We will be using the chi-square, X^2, statistic in this process.

Goodness of Fit – Equal Expected Frequencies

Tina thinks all fast food is the same. In her opinion, one franchise is just as bad as another. Tina also believes that everyone feels the same way. She is sure no one she surveys will express a preference for one fast food vendor over another. She sets out to test her hypothesis by randomly surveying 108 people. She asks them to pick their favorite fast-food provider from a list of three popular establishments. Tina recorded the following results:

	McRonald's	Burger Queen	Chicken Deluxe	Total
Favorite vendor	24	25	59	108

Step 1: State the null and alternate hypothesis:

    H_0 There is no difference in the preference of fast food vendors.

    H_1 There is a difference in the preference of fast food vendors.

Step 2: Select the level of significance:
Tina selects a .05 level of significance, giving her a 5% chance of making a Type I error.

Step 3: Determine the test statistic:
The goodness of fit uses the X^2, chi-square, distribution.

Step 4: Determine the critical value (decision rule):
The degrees of freedom for the X^2 distribution are k – 1. In her survey, Tina has three categories, fast food vendors, so the degrees of freedom are 2.

Locating the .05 right tail area (like the *F* distribution, the X^2 cannot be negative and therefore only has the right tail), with 2 degrees of freedom results in a critical X^2 value of 5.991 as shown in Figure 41.

| | Right tail area | | | |
df	0.1	0.05	0.02	0.01
1	2.706	3.841	5.412	6.635
2	4.605	5.991	7.824	9.210
3	6.251	7.815	9.837	11.345
4	7.779	9.488	11.668	13.277
5	9.236	11.070	13.388	15.086
6	10.645	12.592	15.033	16.812
7	12.017	14.067	16.622	18.475
8	13.362	15.507	18.168	20.090
9	14.684	16.919	19.679	21.666
10	15.987	18.307	21.161	23.209
11	17.275	19.675	22.618	24.725
12	18.549	21.026	24.054	26.217
12	18.549	21.026	24.054	26.217
13	19.812	22.362	25.472	27.688
15	22.307	24.996	28.259	30.578

Figure 41: X^2 dstribution table

Step 5: Take a sample, compute the test statistic and make a decision

The expected frequency is needed to compute the X^2 statistic. In this case, since Tina expected there to be no preferences. So, the frequency expected for each vendor is the same, 36, (108/3).

	Frequency Observed	Frequency Expected
McRonald's	24	36
Burger Queen	25	36
Chicken Deluxe	59	36
Total	108	108

The formula to compute the X^2 statistic is:

$$x^2 = \sum \left[\frac{(f_o - f_e)^2}{f_e} \right]$$

f_o = frequency observed
f_e = frequency expected

	Frequency Observed	Frequency Expected	$(f_o - f_e)^2$	$\frac{(f_o - f_e)^2}{f_e}$
McRonald's	24	36	144	4.000
Burger Queen	25	36	121	3.361
Chicken Deluxe	59	36	529	14.694
Total	108	108		22.056

Tina's computed X^2 statistic is larger than the critical value and is in the rejection area. At a significance level of .05, Tina rejects the null hypothesis which was her belief that there is no preference for one fast food restaurant over another. Tina has learned that perhaps some people do like fast food, or at least prefer the establishments they frequent.

Goodness of Fit – Unequal Expected Frequencies

A recent study conducted by the Chicken Growers of America, found that people prefer chicken sandwiches to burgers on a 2 to 1 ratio. Hearing this, Tina is still not convinced that anyone could really have a preference for any type of fast food. She sets out to use her previously collected data to prove this study wrong.

Step 1: State the null and alternate hypothesis:

H_0 There is no difference between the results reported by the chicken growers and the results of Tina's survey.

H_1 There is a difference between the results reported by the chicken growers and the results of Tina's survey.

Step 2: Select the level of significance:
Tina selects a .02 level of significance, giving her a 2% chance of making a Type I error. She is sure selecting this level of significance will strengthen her campaign to stamp out fast food, if her hypothesis proves to be correct.

Step 3: Determine the test statistic:
The goodness of fit uses the X^2, chi-square, distribution.

Step 4: Determine the critical value (decision rule):
The degrees of freedom for the X^2 distribution are k – 1. In her survey, Tina has three categories, fast food vendors, so the degrees of freedom are 2.

Locating the .02 right tail area with 2 degrees of freedom results in a critical X^2 value of 7.824.

| | Right tail area | | | |
df	0.1	0.05	0.02	0.01
1	2.706	3.841	5.412	6.635
2	4.605	5.991	7.824	9.210
3	6.251	7.815	9.837	11.345
4	7.779	9.488	11.668	13.277
5	9.236	11.070	13.388	15.086
6	10.645	12.592	15.033	16.812
7	12.017	14.067	16.622	18.475
8	13.362	15.507	18.168	20.090
9	14.684	16.919	19.679	21.666
10	15.987	18.307	21.161	23.209
11	17.275	19.675	22.618	24.725
12	18.549	21.026	24.054	26.217
12	18.549	21.026	24.054	26.217
13	19.812	22.362	25.472	27.688
15	22.307	24.996	28.259	30.578

Step 5: Take a sample, compute the test statistic and make a decision

	Frequency Observed	Frequency Expected	$(f_o - f_e)^2$	$\dfrac{(f_o - f_e)^2}{f_e}$
McRonald's	24	21.6	5.76	0.267
Burger Queen	25	21.6	11.56	0.535
Chicken Deluxe	59	43.2	249.64	5.779
Total	108	108		6.581

Tina's computed X^2 statistic, 6.581 is not larger than the critical value, so it is in the "do not reject" area. Once again, Tina's assertions have not proven to be true. She cannot reject the null hypothesis and her research seems to agree with the Chicken Growers data.

Tina gives up on her research and decides to try a sandwich from Chicken Deluxe.

Chapter Highlights

Goodness of fit tests allow us to do hypothesis testing on data that uses the nominal scale of measurement. Goodness of fit tests, which compare actual counts observed to the counts we expected, use the chi-square, X^2.

The formula to compute the X^2 statistic is:

$$x^2 = \sum \left[\frac{(f_o - f_e)^2}{f_e} \right]$$

f_o = frequency observed
f_e = frequency expected

Index

Analysis of Variance 136
ANOVA
    Comparing two population variances . 136
ANOVA table
    creating 138
Binomial Distribution
    Mean 67
    Variance 67
Central Limit Theorem 91
central tendency 24
    Arithmetic Mean 24
    Median 26
    Mode 28
Coefficient of Determination 153
Compliment Rule 52
Confidence interval 98
Contingency Tables 55
Correlation Coefficient 45, 148, 156, 160, 161
Counting principles
    Combination Formula 56
    Multiplication Formula 56
Counting Principles 56
    Permutation Formula 57
Discrete Probability Distribution
    Mean 63
    Variance 64
dispersion - measures of 28
    Standard Deviation 34
    Variance 30
disperson - measures of
    range 29
Dummy Variables 154
Empirical Rule 35, 104
Five step hypothesis testing procedure ... 108
Frequency distributions 14
    quantitative data 17
Goodness of fit 164
hypothesis test
    dependent samples 128
    One sample 112
    one sample - unknown population standard deviation 115
    Two sample - proportions 127
Hypothesis test
    one proportion 117
    two sample 122
    Two Sample– unknown standard deviation 124
Intercept 150
Margin of Error
    Known standard deviation 98
    Unkown population standard deviation 100
measurement scales 9
    interval 9
    nominal 9
    ordinal 9
    ratio 10
Multiple regression 153
Normal Probability Distributions 78
Parameters vs. Statistics 25
Point Estimate 98
population 5
Position - measures of 40
Probability
    classical 51
    empirical 51
    Rules of Addition 53
    subjective 51
Probability Distributions
    Binomial 65
    Continuous 76
    Discrete 62
    Normal 78
    Poisson 69
Proportions
    Confidence intervals for 102
Regression Analysis 150
Regression formula 151
Relative frequencies 16
    cumulative 17
sample 6
sample size
    choosing 103
Sampling distribution of the sample mean 89

Sampling error 88
Sampling methods.................................... 88
    Simple Random.................................... 88
    Systematic Random 88
Scatter Plot... 44
Skewness............................... 41, 42, 43, 47
Slope 150
Standard Error of the estimate 153
Standard error of the mean....................... 94
statistics
    defined... 4

    descriptive 5
    inferential 5
Uniform Distributions............................... 76
Variables 7
    Continuous 8
    Dependent and Independent................ 148
    Discrete 8
    qualitative.. 7
    Quantitative... 7
Z score
    Sample mean... 92

www.ingramcontent.com/pod-product-compliance
Lightning Source LLC
Chambersburg PA
CBHW060414220526
45465CB00008B/2882